Nanotechnology

FOR

DUMMIES®

2ND EDITION

Nanotechnology
FOR
DUMMIES®
2ND EDITION

by Earl Boysen and Nancy Boysen

Foreword by Desiree Dudley and Christine Peterson
Foresight Institute

WILEY

Wiley Publishing, Inc.

Nanotechnology For Dummies®, 2nd Edition

Published by
Wiley Publishing, Inc.
111 River Street
Hoboken, NJ 07030-5774

www.wiley.com

Copyright © 2011 by Wiley Publishing, Inc., Indianapolis, Indiana

Published by Wiley Publishing, Inc., Indianapolis, Indiana

Published simultaneously in Canada

For general information on our other products and services, please contact our Customer Care Department within the U.S. at 877-762-2974, outside the U.S. at 317-572-3993, or fax 317-572-4002.

For technical support, please visit www.wiley.com/techsupport.

Wiley also publishes its books in a variety of electronic formats and by print-on-demand. Not all content that is available in standard print versions of this book may appear or be packaged in all book formats. If you have purchased a version of this book that did not include media that is referenced by or accompanies a standard print version, you may request this media by visiting http://booksupport.wiley.com. For more information about Wiley products, visit us www.wiley.com.

Library of Congress Control Number: 2011932269

ISBN: 978-0-470-89191-9 (pbk); ISBN: 978-1-118-13686-7 (ebk); ISBN: 978-1-118-13687-4 (ebk); ISBN: 978-1-118-13688-1 (ebk)

Manufactured in the United States of America

10 9 8 7 6 5 4 3 2 1

WILEY

About the Authors

Earl Boysen spent 20 years as an engineer in the semiconductor industry and runs two web sites, UnderstandingNano.com and BuildingGadgets.com. Earl holds a Masters in Engineering Physics from the University of Virginia. He was coauthor of the first edition of *Nanotechnology For Dummies* and *Electronics For Dummies*. He also coauthored *The All New Electronics Self-Study Guide* from Wiley Publishing.

Nancy Boysen is the author of more than 60 books on technology topics (under the name Nancy Muir), including *Microsoft Project For Dummies* and *iPad All-In-One For Dummies,* and contributed to the college textbook *Our Digital World* from Paradigm Publishing. She is the senior editor for UnderstandingNano.com and runs two other web sites, TechSmartSenior.com and iPadMadeClear.com.

Dedication

To Nettie Boysen, Earl's mom, for providing the love and support that helped him to follow his dreams.

Authors' Acknowledgments

The authors wish to thank Katie Feltman for hiring them to write this edition of *Nanotechnology For Dummies*. Also thanks to Susan Pink for leading the way as project editor, and Lisa Reece for her excellent technical edit. We also want to express our gratitude to colleagues in the world of nanotechnology who have allowed us to use their artwork in the book, and who have shared their expertise generously during our research. Finally, sincere thanks to Desiree Dudley and Christine Peterson of Foresight Institute for contributing the book's foreword.

Publisher's Acknowledgments

We're proud of this book; please send us your comments at http://dummies.custhelp.com. For other comments, please contact our Customer Care Department within the U.S. at 877-762-2974, outside the U.S. at 317-572-3993, or fax 317-572-4002.

Some of the people who helped bring this book to market include the following:

Acquisitions and Editorial

Project Editor: Susan Pink

Acquisitions Editor: Katie Feltman

Copy Editor: Susan Pink

Technical Editor: Lisa Reece

Editorial Manager: Jodi Jensen

Media Development Project Manager: Laura Moss-Hollister

Media Development Assistant Project Manager: Jenny Swisher

Media Development Associate Producers: Josh Frank, Marilyn Hummel, Douglas Kuhn, and Shawn Patrick

Editorial Assistant: Amanda Graham

Sr. Editorial Assistant: Cherie Case

Cartoons: Rich Tennant (www.the5thwave.com)

Composition Services

Project Coordinator: Sheree Montgomery

Layout and Graphics: Lavonne Roberts, Kim Tabor

Proofreader: Laura Bowman

Indexer: Potomac Indexing, LLC

Special Help
Karl Brandt, Melissa Smith, and Shawn Frazier

Publishing and Editorial for Technology Dummies

    Richard Swadley, Vice President and Executive Group Publisher

    Andy Cummings, Vice President and Publisher

    Mary Bednarek, Executive Acquisitions Director

    Mary C. Corder, Editorial Director

Publishing for Consumer Dummies

    Kathy Nebenhaus, Vice President and Executive Publisher

Composition Services

    Debbie Stailey, Director of Composition Services

Contents at a Glance

Table of Contents

· ·

Foreword

. . . .

Realizing the Potential of Nanotechnology

*W*hat is nanotechnology? It's a big word: tiny in scale but infinitely immense in possibility. In the Silicon Valley era of tech bubbles and busts, you may have heard nanotechnology bandied about as the new thing, along with biotech, artificial intelligence, private space travel, and more.

But what does nanotechnology mean? Perhaps the most influential early reference to the field we now call nanotechnology was on December 29, 1959. That evening, one of the most famous and beloved physicists of all time, Richard Feynman, gave a dinner lecture at the California Institute of Technology entitled "There's Plenty of Room at the Bottom," where he discussed the potential in our increasing knowledge and ability to manipulate matter:

> *The principles of physics, as far as I can see, do not speak*
> *against the possibility of maneuvering things atom by atom . . .*
> *Put atoms down where the chemist says, and so you make*
> *the substance.*

Feynman's visionary forecast was before its time; however, excitement about the field truly began to manifest with the invention of the scanning tunneling microscope (STM) by Gerd Binnig and Heinrich Rohrer of IBM in 1981, and the field's first book, *Engines of Creation: The Coming Era of Nanotechnology,* written in 1986 by K. Eric Drexler. That year, Drexler and Christine Peterson formed the Foresight Institute, a nonprofit think-tank whose purpose is to advance the ethical development of beneficial nanotechnology.

Twenty-five years later, the field has blossomed. Billions of dollars go into nanotechnology research and development every year. More than a hundred major academic institutions, governmental organizations, research facilities, and advocacy groups in the world are dedicated to nanotechnology. We can see cells, atoms, and DNA at the sub-nanometer level with scanning electron and tunneling microscopes, measure and move molecules with atomic force and probe microscopes, "paint" with molecules using dip-pen lithography, and even snip and modify DNA using manmade DNA "walkers." We have begun putting the first labs on chips, identifying and even killing cancer cells with nanoscale techniques.

We have come so far. But have we reached a truly nanoscale control of matter? As so often happens, humanity has found that the devil is in the details: realizing the dream of molecular- and atomic-level precision is more difficult than its conception. Quantum physics and its mechanical effects become much more important on the nanoscale, and our understanding of the laws of nature at this scale is advancing but by no means fully comprehensive. Even with all our advances to date, processes for building truly precise three-dimensional structures through molecular manufacturing are still in-progress.

Lacking truly accurate understanding and precise application, media and industry have capitalized on the hopes and fears of a naive society fascinated by the potential in health, life extension, space travel, and green energy. *Nanotechnology* has become a much-hyped magical buzzword that glamorizes — or demonizes — today's production of imprecise nanoscale blobs.

However, despite real limitations, microscale and nanoscale progress to date is still impressive. The Information Age completely transformed our world by controlling those "blobs" of matter on a micronscale; the average cellular phone in your pocket today has more processing power than machines that filled entire basements in the 1980s. Articles and books that could take months to find can now be downloaded in moments; family members can call their loved ones from remote areas around the world; 911 emergency services can be at your car accident far faster than ever thought possible. Information sharing, communication, and real knowledge propagation that took weeks or months — or even years — can now be achieved faster than ever before because human beings had the courage to understand, develop, and implement new knowledge and technologies.

But this world-changing progress is merely the microscale; time has already started to show that we can do better. And even more is possible. Imagine a world in which a family of four can take a trip to the moon for the price of a Sunday drive — because the materials and fuel are so light, strong, and inexpensively made. Imagine a world in which nanoscale devices can go in and help rebuild your grandmother's heart, or your own arteries. A world in which chemical pollution no longer exists.

You may think "that sounds like science fiction." Well, that it is. In 1995, best-selling author Neal Stephenson wrote about this kind of world in a Hugo Award-winning book called *The Diamond Age*. And that world is truly a different world than the one we live in now. But this kind of grand, forward-thinking vision has always inspired human progress. Ideas are first whispered or hastily scrawled by those starry-eyed dreamers who dare to imagine something more, something better. In the history of human civilization, the curious inventors, the doers, the makers, and the courageous leaders are the ones who dare to try, to understand, to be inspired, to create, to build: to take those far-off dreams and make them real. The road to truly great dreams is often a long one, and humanity almost always takes more time, energy, work, and earnest collaboration than imagined to fully build and travel this road, — especially to travel it well.

Nanotechnology For Dummies, 2nd Edition, guides the reader through a bright path of progress and possibility, on a road that will eventually lead to all that nanotechnology promises. This book also serves as an entrée into the basic concepts, achievements, problems, and prospects in this exciting field. We hope the knowledge will inspire you to help us create a better world.

— Desiree Dudley and Christine Peterson
— Foresight Institute

Introduction

· ·

*I*f you are one of the many who has read headlines about nanotechnology and the incredible things it is making possible in our world, you've probably bought this book to find out what the fuss is all about. Nanotechnology has been touted as both a Holy Grail of science that can cure all ills and a dangerous manipulation of matter that could cause the end of our world. So just what is nanotechnology and what could it make possible?

Nanotechnology For Dummies, 2nd Edition, helps you get a good grounding in nanotechnology history, concepts, and applications while clearing up some of the hype. As you work your way through its chapters, you will discover some fascinating facts about nanotechnology past, present, and future.

About This Book

Nanotechnology is probably the most promising branch of science today. It holds out the possibility of clean air, cheap energy, and longer life. In fact, almost every industry today is using or considering nano for their business, and most countries have some level of nanotechnology research and development.

Although nanotechnology can be a complex topic, we've made every effort to give you a good overview of its many aspects while not driving you to distraction with jargon and technical talk. We explore not only the concepts behind nanotechnology but also how it's being applied in the real world. We even take several glimpses into the future to explain all the things that nano may help make possible.

If you are looking at nanotechnology as a career, an investment opportunity, or a scientific field of study, or are just curious, this book will provide you with answers.

Finally, because nanotechnology is a fast-moving field, note that we provide updated information for our readers on our web site, www. UnderstandingNano.com.

Foolish Assumptions

While writing this book, we assumed that you would have at least a passing interest in science, but we didn't assume they you are a scientist. We therefore ried to put things in simple terms and define technical terms when they first appear as well as in a glossary.

We also assumed that most of you have access to the Internet, so we've included throughout the book URLs of sites you may want to visit for more information or updates. In case you would prefer not to type the URLs to access these sites, we've provided links to each site on our web site at www.UnderstandingNano.com/nanotechnology-links.html.

Finally, we assumed that you want to go right to the information that's most useful to you, so we wrote this book in a way that doesn't require you to read it in any particular order. Jump in wherever you like!

How This Book Is Organized

This book is conveniently divided into several handy parts to help you find the information you need.

Part I: Nanotechnology Basics

The chapters in Part I introduce you to nanotechnology: what it is, where it came from, and the people who made key discoveries to advance the science. We also include chapters about nano materials, techniques used in manipulating those materials, and tools that every nanotechnologist should have in his or her nano toolkit.

Part II: Nano Applications

Nanotechnology is a science that has applications in almost every area of life, from health care to manufacturing, space travel to improving our environment. In Part II, we explore what's being done, developed, or just imagined in various industries and settings.

Part III: Nanotechnology and People

Nanotechnology may be relevant to you in a few keys ways. In Part III, we explore the ethical, safety, and regulatory issues that may have an effect on how you interact with nanotechnology products or processes in your daily life. We also explore the educational and career opportunities you might want to take advantage of to become part of this fascinating field.

Part IV: The Part of Tens

The field of nanotechnology has many players and many resources. In the three chapters in Part IV, we offer an overview of ten great web sites related to nano, ten interesting nanotechnologies, and ten research labs at the forefront of nanotechnology research and development.

Glossary

The glossary puts all the nanotechnology terms we introduce in the book in one spot to give you a handy, alphabetical reference.

Icons Used in This Book

This book, like all *For Dummies* books, has little icons in the margin. When you see one of the following icons, take heed.

This icon indicates some fascinating statistic or fact about nano that curious readers may want to explore.

Look for this icon to get suggestions of books, web sites, videos, and other sources for additional information on some aspect of nanotechnology.

When updates to a topic may be posted on the online companion web site, this icon reminds you to go there to get the latest info.

This icon points out information that isn't vital to understanding the topic under discussion but provides interesting technical background or explanations for the more scientifically inclined reader.

Where to Go from Here

Nanotechnology is a fascinating field that holds out promise for our future welfare and well-being. The developments in nano are fascinating, and understanding them provides you with some interesting scientific knowledge in several areas, including physics and chemistry. Dive into this book in whatever fashion you please. If you need to get a grounding in nano basics, start at Chapter 1 and work your way through. Or if a particular topic interests you, use the table of contents to find the chapter that discusses that topic and jump right in. In either case, enjoy the information provided in the chapters of this book as you become one of the nano-savvy.

Part I
Nanotechnology Basics

The 5th Wave   By Rich Tennant

NANOTECHNOLOGY
IN COMPUTER DEVELOPMENT

"Just don't sneeze again until we locate the servers, storage device, and mainframes."

In this part . . .

This part starts from square one, explaining all the nanotechnology basics. The chapters here introduce you to nanotechnology: what it is, where it came from, and the people who made key discoveries to advance the science over the last several decades.

Next, we tell you all about the building blocks of nanotechnology. You find out about nanomaterials, techniques used in manipulating those materials, and tools that every nanotechnologist should have in his or her nano toolbox.

If you're new to nanotechnology, this part gives you a great grounding in everything nano.

Chapter 1

Introduction to Nanotechnology Concepts

In This Chapter

▶ Exploring the definition of nanotechnology

▶ Understanding how nano-sized materials vary from bulk materials

▶ Examining the bottom-up and top-down approaches to nano

▶ Following nano's role across disciplines and industries

Nanotechnology has been around as a recognized branch of science for only about fifty years, so it's a baby compared to physics or biology, whose roots go back more than a thousand years. Because of the young age of nanotechnology and our still-evolving understanding of it, defining it is an ongoing process, as you find in this chapter.

In addition, we help you understand nano by comparing it to more familiar concepts, such as atomic structure, and look at how materials change at the nano level.

Finally, the promise nanotechnology holds for the human race ranges from extending our lives by centuries to providing cheap energy and cleaning our air and water. In this chapter, you explore the broad reach that nanotechnology has across several scientific disciplines and many industries.

What Is Nanotechnology, Anyway?

To help you understand exactly what nanotechnology is, we start by providing a definition — or two. Then we explore how nano-sized particles compare with atoms.

Pinning down a definition

Nanotechnology is still evolving, and there doesn't seem to be one definition that everybody agrees on. We know that nano deals with matter on a very small scale: larger than atoms but smaller than a breadcrumb. We know that matter at the nano scale can behave differently than bulk matter. Beyond that, individuals and groups focus on different aspects of nanotechnology as a discipline. Here are a few definitions of nanotechnology for your consideration.

The following definition is probably the most barebones and generally agreed upon:

> *Nanotechnology is the study and use of structures between 1 nanometer (nm) and 100 nanometers in size.*

To put these measurements in perspective, you would have to stack 1 billion nanometer-sized particles on top of each other to reach the height of a 1-meter-high (about 3-feet 3-inches-high) hall table. Another popular comparison is that you can fit about 80,000 nanometers in the width of a single human hair.

The word *nano* is a scientific prefix that stands for 10^{-9} or 1 billionth; the word itself comes from the Greek word *nanos,* meaning dwarf.

The next definition is from the Foresight Institute and adds a mention of the various fields of science that come into play with nanotechnology:

> *Structures, devices, and systems having novel properties and functions due to the arrangement of their atoms on the 1 to 100 nanometer scale. Many fields of endeavor contribute to nanotechnology, including molecular physics, materials science, chemistry, biology, computer science, electrical engineering, and mechanical engineering.*

The European Commission offers the following definition, which both repeats the fact mentioned in the previous definition that materials at the nanoscale have novel properties, and positions nano vis-à-vis its potential in the economic marketplace:

> *Nanotechnology is the study of phenomena and fine-tuning of materials at atomic, molecular and macromolecular scales, where properties differ significantly from those at a larger scale. Products based on nanotechnology are already in use and analysts expect markets to grow by hundreds of billions of euros during this decade.*

This next definition from the National Nanotechnology Initiative adds the fact that nanotechnology involves certain activities, such as measuring and manipulating nanoscale matter:

Nanotechnology is the understanding and control of matter at dimensions between approximately 1 and 100 nanometers, where unique phenomena enable novel applications. Encompassing nanoscale science, engineering, and technology, nanotechnology involves imaging, measuring, modeling, and manipulating matter at this length scale.

The last definition is from Thomas Theis, director of physical sciences at the IBM Watson Research Center. It offers a broader and interesting perspective of the role and value of nanotechnology in our world:

[Nanotechnology is] an upcoming economic, business, and social phenomenon. Nano-advocates argue it will revolutionize the way we live, work and communicate.

Before nano there was the atom

If you remember your high school science class, you know something about atoms, so we'll take that as our starting point in explaining the evolution of nanotechnology. Figure 1-1 is an illustration of an atom containing positively charged protons and neutral neutrons in the nucleus (center) of the atom, as well as negatively charged electrons in orbit around the nucleus.

Figure 1-1: Simple model for the structure of an atom.

The word *atom* comes from the Greek word for indivisible, *atomos.* The atomic bomb demonstrated that atoms can indeed be split, but way back in 450 B.C. they were blissfully unaware of such possibilities. In 1803 John Dalton discovered that elements such as water are actually collections of atoms. These collections, called molecules, have different characteristics from the separate atoms (think of two hydrogen atoms combining with one oxygen atom and the wet result of H_2O).

Today we recognize that some of Dalton's original theory of the atom doesn't hold water. Still, the most important concepts, that chemical reactions involve the joining and separating of atoms and that atoms have unique properties, are the basis of today's physical science.

The idea that atoms combine to form molecules such as water is key to chemistry, biology, and nanotechnology. The work of Dalton and many other scientists has allowed chemists to develop useful materials, such as plastics, as well as destructive materials, such as explosives. All bulk materials are made up of atoms, so it was necessary to first understand atoms to learn how to make new materials. Scientists could draw conclusions about atoms based on the properties of the materials they produced, even though they couldn't see inside an atom.

An important point to underscore is that nobody has ever seen the structure of an atom. Even today's most sophisticated microscopes don't reveal the details of atoms, just fuzzy pictures of tiny orbs. All the information about the structure of atoms is based on empirical evidence. Scientists determined that each type of atom absorbed different frequencies of light and then used those differences to make a model of the structure of electrons around the nucleus of each atom. Other scientists bombarded atoms with very small high-energy particles and analyzed what type of particles resulted from collisions with the atomic nucleus to guess at what was inside the nucleus of each type of atom. Then scientists did the math and developed a model of each atom to match their results. The way we describe atoms to our high school students today continues to evolve as physicists probe atoms with higher and higher energy particles to provide more details about the components of the atomic nucleus.

So how does all this information about atoms relate to nanotechnology? *Nanoparticles* (particles whose diameter, width, or length is between 1 nanometer and 100 nanometers) are bigger than atoms and, like atoms, are around us everyday. They are given off by candle flames, wood fires, diesel engines, laser printers, vacuum cleaners, and many other sources. Scientists worked with nanoparticles for centuries before these particles had a name. But unlike atoms — and this is a big difference — we can now see the structure of nanoparticles. This breakthrough came a few decades ago with the advent of electron microscopes. Figure 1-2 shows the structure of some key nanoparticles (such as the DNA molecule in the bottom left) and their size in relation to other materials.

With our understanding of atomic theory and the ability to see things at the nanoscale, we now have knowledge in place that allows us to manipulate matter in ways never before possible.

See Chapter 3 for more about nanoparticles and materials and Chapter 4 for information about how these can be manipulated.

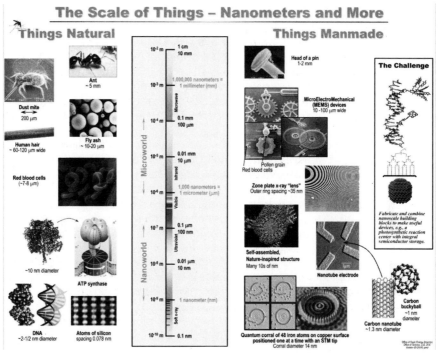

Figure 1-2:
The scale of things.

Moving from half-baked to Bakelite

Have you ever wondered what the first entirely man-made (synthetic) substance was? Turns out it was a resin-based material called Bakelite, the very first plastic. Bakelite was developed by a chemist named Leo Baekeland. Producing Bakelite involves the application of heat and pressure while mixing two chemicals. Bakelite hardens to form the shape of any mold you pour it into. Its unique composition means that you can't burn it or dissolve it with any commonly available acid. The United States military saw Bakelite as the basis for producing lightweight weapons, and Bakelite became a part of most weapons used in World War II.

Bakelite also found its way into products such as electrical insulators and dishware. The stuff doesn't break, crack, or fade. (Okay, this sounds like an infomercial, but this stuff is really amazing.) Today other plastics have largely replaced Bakelite in general use, but it was a big breakthrough in its time.

Approaching Nanotechnology from Above and Below

How we should use our knowledge of nanoparticles has been a subject of much debate. Nanotechnologists have offered two approaches for fabricating materials or manipulating devices using nanotechnology: top down and bottom up.

Imagine you need to build the tiniest computer chip possible. Using the *bottom-up* approach, you would use nanotechnology to assemble the chip atom by atom, placing each type of atom in a specific location to build the circuit. With the *top-down* approach, you would instead create the computer chip by carving away at bulk material — much like a sculptor and his artwork — to create nano-sized features, never dealing with the atomic level of matter.

The top-down method is currently in use to manufacture computer chips as well as other products you use every day. The bottom-up method is in the theoretical stage, with researchers doing initial experiments to develop these techniques.

Nanotechnologists also use a technique called self-assembly, discussed in Chapter 4, to build structures using nanoparticles. *Self-assembly* involves creating conditions such that atoms and molecules arrange themselves in a specific way to create a material. Some consider this one form of the bottom-up approach.

Examining four generations of nano development

Mihail (Mike) Roco of the Nanotechnology Initiative has suggested that nanotechnology development will occur in four generations. Currently, according to Roco, we're in an era of passive nanostructures, which he describes as "materials designed to perform one task." In the second generation, which we have already entered, we are using "active nanostructures for multitasking." These nanostructures would include devices to deliver drugs in a targeted way. The third generation would include nanosystems that might involve thousands of components interacting with each other. Finally, several years in the future, we may see integrated nanosystems, including systems within systems that could accomplish far more than we can today, such as sophisticated molecular manufacturing of genes inside the DNA of targeted cells and nanosurgery for healing wounds on the cellular level.

Understanding How Nano Changes Things

We've stated that materials at the nano level have different characteristics than so-called bulk materials. That's a key concept worthy of a little more explanation.

Nanoparticles are so small they contain just a few atoms to a few thousand atoms, as opposed to bulk materials that might contain many billions of atoms. This difference is what causes nano materials to have unique characteristics, including how they react with other materials, their color, and even how they melt at high temperatures.

In this section, we explore these differences and help you begin to understand the changes they make possible.

Reacting with other elements

One aspect of how nano-sized particles act differently is how they behave in chemical reactions. One of the most interesting examples of this involves gold.

Gold is considered an inert material in that it doesn't corrode or tarnish (which is why you paid so much for that ring on your finger). Normally, gold would be a silly material to use as a catalyst for chemical reactions because it doesn't do much. However, break gold down to nanosize (approximately 5 nanometers) and it can act as a catalyst that can do things like oxidizing carbon monoxide.

This transformation works as follows. The smaller the nanoparticle, the larger the proportion of atoms at the surface, and the larger proportion of atoms at the corners of the crystal. While in the bulk form, each gold atom (except the small percentage of them at the surface) is surrounded by twelve other gold atoms; even the gold atoms at the surface have six adjacent gold atoms. In a gold nanoparticle a much larger percentage of gold atoms sits at the surface. Because gold forms crystalline shapes, as shown in Figure 1-3, gold atoms at the corners of the crystals are surrounded by fewer gold atoms than those in the surface of bulk gold. The exposed atoms at the corners of the crystal are more reactive than gold atoms in the bulk form, which allows the gold nanoparticles to catalyze reactions.

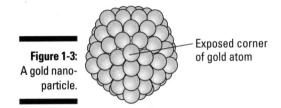

Figure 1-3:
A gold nano-
particle.

Exposed corner
of gold atom

Changing color

It turns out that gold's capability to catalyze reactions is not the only thing that changes at the nanoscale. Gold can actually change color depending on the size of the gold particles.

One of the characteristics of metals is that they are shiny because light reflects off their surfaces. This reflectivity has to do with electron clouds at the surface of metals. Because photons of light can't get through these clouds and therefore aren't absorbed by the electrons bound to atoms in metals, the photons are reflected back to your eye and you see that shiny bling quality.

In addition to this shininess, metals have different colors because different colors are reflected more strongly than others. Gold and copper, for example, have lower reflectivity for shorter wavelengths, such as blue, than for longer wavelengths, such as yellow, green, and red, causing a gold tone. Silver has a more constant reflectivity across wavelengths, so it reflects all colors, making it seem more like an absence of color (white).

In bulk form, gold reflects light. At the nanoscale, the electron cloud at the surface of a gold nanoparticle resonates with different wavelengths of light depending upon their frequency. Depending on the size of the nanoparticle, the electron cloud will be in resonance with a particular wavelength of light and absorb that wavelength. A nanoparticle of about 90 nm in size will absorb colors on the red and yellow end of the color spectrum, making the nanoparticle appear blue-green. A smaller-sized particle, about 30 nm in size, absorbs blues and greens, resulting in a red appearance.

Nanotechnologists are debating the possible use of this color-changing characteristic to build sensors in fields such as medicine.

Melting at lower temperatures

Another characteristic that varies at the nano level is the temperature at which a material melts. In bulk form, a material, such as gold, has a certain melting temperature regardless of whether you're melting a small ring or a bar of gold. However, when you get down to the nanoscale, melting temperatures begin to vary by as much as hundreds of degrees.

Gold nanoparticles melt at relatively low temperatures (~300 degrees celsius for 2.5 nm size) whereas a slab of gold melts at a toasty 1064 °C.

This difference in melting temperature again relates to the number of atoms on the surface and corners of gold nanoparticles. With a greater number of atoms exposed, heat can break down the bond between them and surrounding atoms at a lower temperature. The smaller the particle, the lower its melting point.

Nano Is Everywhere

Nanotechnology is sometimes referred to as a general-purpose technology because in its more advanced stages it will have a significant impact on almost all industries and all areas of society.

Nanotechnology is unlike other scientific disciplines you may be familiar with in its breadth. It pulls in information from physics, chemistry, engineering, and biology to study and use materials at the nano level to achieve various results.

It turns out that being able to see and work with things on a very small level has some very big ramifications not isolated to one industry or field. In this section you get an idea of the types of applications and future changes nano is making possible.

Applying nano in various settings

Nanotechniques can be applied in many different settings and many different applications. It crosses industries, enabling achievements in areas such as manufacturing, medicine, space travel, energy, and the environment. The techniques being developed for creating and manipulating particles at the nano level hold out the hope for curing diseases such as cancer, cleaning our air, and producing cheap energy.

Think of some other scientific disciplines. Medicine pertains only to healthcare; astronomy is lost in the stars; zoologists focus on animals. But nanotechnology doesn't have that single focus; in that respect, it's more like physics or chemistry, scientific disciplines whose discoveries can be used in many areas and many industries and with many other sciences.

This range of effect is indicated by the many federal agencies that participate in the *National Nanotechnology Initiative,* whose web site is shown in Figure 1-4. These agencies include the Departments of Agriculture, Defense, Energy, Labor, Transportation, and Treasury, as well as organizations such as the National Institute of Health, Forest Service, NASA, and the Environmental Protection Agency.

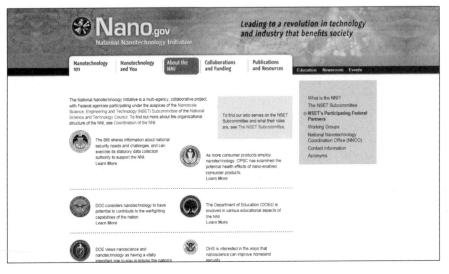

Figure 1-4:
Some of the agencies participating in the National Nano-technology Initiative.

Nano is also used in many commercial settings, many of which you'll hear about in more detail in subsequent chapters. For example, nano materials are used to

- Add strength to materials used in products ranging from tennis rackets to windmills
- Act as catalysts in chemical manufacturing
- Help absorption of drugs into the body
- Add stain resistance to fabrics used in clothing
- Make medical imaging tools such as MRIs function more accurately
- Improve the efficiency of energy sources such as batteries and fuel cells
- Purify drinking water and clean up our air

There may even be nanoceramics in your dental implants, taking advantage of the fact that their properties can be adjusted to match the properties of the tissue surrounding them. And just about every electronic gadget you own probably has some type of *nanomaterial* in it, especially in the chips used in computing devices.

Taking a clue from educators

If you want to find out whether a field involves various disciplines, check out the university programs offered for those interested in the field. What you'll find is that nanotechnology is such an interdisciplinary field that many educational programs accept students from a range of disciplines.

Mahbub Uddin and Raj Chodhury, in a conference paper on Nanotechnology Education, make this statement about the challenge of providing nanotechnology education: "Nanotechnology is truly interdisciplinary. An interdisciplinary curriculum that encompasses a broad understanding of basic sciences intertwined with engineering sciences and information sciences pertinent to nanotechnology is essential."

If you'd like to read the full article, go to this URL: `www.actionbioscience.org/education/uddin_chowdhury.html`.

Various programs at institutions such as Northeastern University, Rice University (see Figure 1-5), and Penn State offer courses and degrees in everything from nanomedicine to nanobusiness. These programs are available to students majoring in such disciplines as Biology, Chemistry, Physics, Chemical Engineering, Mechanical/Industrial Engineering, Electrical/Computer Engineering, Pharmaceutical Sciences, Materials Science and Engineering, Integrative Biosciences, and Applied Chemistry.

It's clear from these examples that preparing students for a career in nanotechnology can involve knowledge of both nanotechnology and the specialty of their choosing.

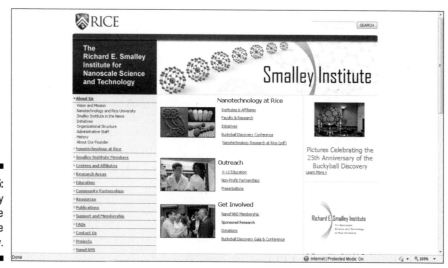

Figure 1-5:
The Smalley
Institute
at Rice
University.

Chapter 2

Who's Doing What?

In This Chapter

▶ Examining the evolution of nanotechnology

▶ Exploring the roles people have played in nano

▶ Laying out how companies, research labs, and governments are involved

▶ Surveying what applications are a reality today and what's in the future

*U*nderstanding the concept that nanotechnology deals with materials and their manipulation at a *very small scale* gives you just a basic definition of nanotechnology. However, reviewing both the timeline of discoveries that brought us to our current understanding as well as what individuals, universities, companies, and researchers are doing in the field of nanotechnology will help you get the big picture.

In this chapter we cover the evolution of nanotechnology and some of the key players in that evolution; the who's who of organizations involved in the field of nanotechnology; and some of the products and techniques nanotechnology is making possible today and will make possible in the future.

Understanding the Evolution of Nanotechnology

Although the use of nanotechnology in the form of nanoparticles is ancient, nanotechnology as a field of science appeared on the scene only in the past 80 years or so. In this section we explore some of the key moments and people who have shaped nanotechnology as we know it today.

Viewing the timeline

Nanoparticles have been used by people for thousands of years, but they didn't know it. For example, researchers report that a lead-based hair dye used in ancient Egypt created nanocrystals during the chemical reaction involved in the dyeing process. As early as the tenth century, many European church windows got their color from gold nanoparticles embedded in glass.

The colors occurred because gold nanoparticles exhibit different colors from bulk gold. See Chapter 1 for a discussion of how and why gold at the nanoscale changes color.

But having materials that are all around us behave in a certain way does not a science make. Formalizing this scientific field of study took the awareness of what those particles are and how they behave. And, in the case of nano-technology, being able to see those particles was the key to unlocking the huge potential of the nanoscale world.

Beginning to see things at the nanoscale

The first big breakthrough in seeing nanoparticles came in 1931, when German scientists Ernst Ruska and Max Knoll built the first *transmission electron microscope,* or TEM. This device has several hundred times the power of a traditional microscope, which uses light for magnification. A TEM uses a focused beam of electrons, instead of light, to pass through and magnify an object by a factor of up to about 1,000,000.

In the 1930s scientists had reached the limitations of light microscopes, which can resolve only objects greater than about 200 nm, which is slightly less than the wavelength of visible light. They needed something that would allow them to view structures such as the interior of organic cells, which require the reso-lution of structures of only a few nanometers in size. To be able to view such objects, they worked to develop the TEM.

The transmission electron microscope and its subsequent improvements enabled researchers to see the structure of nanoscale objects. They could now explore the structure of organic molecules that make up the human body, such as proteins, and inorganic materials, such as metals, by examin-ing a cross section of a sample. This capability made possible advances in areas ranging from electronics and medicine to manufacturing. But these advances weren't yet collectively defined as nanotechnology.

Introducing nano to the world: Richard Feynman's role

Some consider that the general concept of nanotechnology started with a talk that Richard P. Feynman, an American theoretical physicist, gave in 1959 at the California Institute of Technology (CalTech). In that talk, titled "There's Plenty of Room at the Bottom," Feynman offered up the possibility of working with and controlling the atoms and molecules that make up matter, which is essentially the bottom-up approach to nanotechnology discussed in Chapter 1.

Feynman was no one-trick pony; he also essentially reinvented quantum electrodynamics, which relates to how light and matter interact at the atomic level. His work changed the world's understanding of how light is absorbed and emitted by electrons and how light is involved when electrons repel each other. Feynman was co-awarded the Nobel Prize for Physics in 1965 for this work.

People have debated whether Feynman was really instrumental in bringing the idea of working with matter at the nanoscale to the forefront. The scientific community didn't seem to take much notice of his talk in the 20 years or so after he gave it. It wasn't until a seminal work on nanotechnology, *Engines of Creation* by Eric Drexler, came out in 1986 that Feynman's talk received this kind of credit. However, in retrospect, Feynman's speech, in which he asked, "What would happen if we could arrange the atoms one by one the way we want them?", did a good job of stating the possibilities of the nanoscale world, and so it marks a notable point in nanotechnology's timeline. In addition, Feynman's interest in nanotechnology and his stature in the scientific community played a role in securing nanotechnology funding years after his talk.

Exploring the role of the scanning tunneling microscope

After people could see objects at the nanoscale and the new science had been identified, the next step in nanotechnology's evolution was to try to move individual atoms. That breakthrough came in 1981 when Gerd Binning and Heinrich Rohrer of IBM Zurich invented a machine named the scanning tunneling microscope (STM), which can move objects at the atomic scale.

And yes, in case you're wondering, the invention of the STM earned Binning and Rohrer the Nobel Prize in Physics in 1986. Are you sensing a pattern here? Because nanotechnology is a field that holds tremendous promise, those involved in this field often merit the scientific community's attention in a big way.

Here's how an STM works. An incredibly sharp metal wire tip (we're talking one or two atoms wide) moves over a surface. When you apply electrical voltage between the tip and the sample it's moving over, you can create an image of the electrical topography of the surface. That image is created based on changes in current that flows from the tip to the surface. Using this method you can get images of matter as tiny as an atom.

But this gets really interesting when you realize that an STM can not only image atoms but also move them. A famous demonstration of this came in 1989 when Don Eigler of IBM arranged 35 atoms on a surface made of nickel to spell out IBM. (We're betting he received a great Christmas bonus that year.)

Moving atoms around doesn't just happen in your average kitchen. This particular demonstration involved the use of a high vacuum and a super-cooled environment at the temperature of liquid helium.

Pursuing buckyballs

In 1985, between the introduction of the STM and the famous demonstration of how it could arrange atoms, a British chemist named Harry Kroto note that chains of carbon atoms were present trillions of kilometers away in space. Kroto conjectured that these chains might have been created in the atmosphere of red giant stars.

Around this time, Kroto connected with Richard Smalley and Robert Curl, American researchers at Rice University who were studying clusters of atoms that were generated when they vaporized metal or semiconductor samples. This trio got together when Kroto came to the states to use Rice University's high-end equipment. To replicate the really hot conditions that exist in the atmosphere of a red giant star (way hotter than Houston in August), they vaporized graphite using a laser in an atmosphere of helium gas.

The procedure produced carbon molecules nobody had ever observed before. They noted that the most common molecule contained 60 carbon atoms. Because the molecules seemed stable (they retained their shape and size), Kroto, Smalley, and Curl guessed that they were spherical, because spherical molecules tend to be more stable. The three scientists finally determined that combining 60 carbon atoms in a spherical shape required interlocking hexagons and pentagons, like those shown in Figure 2-1.

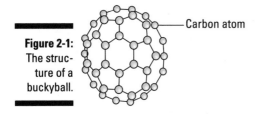

Figure 2-1:
The structure of a buckyball.

Carbon atom

They named this structure *buckyball,* after the designer of the geodesic dome, Buckminster Fuller, because the structure resembled the design of his dome. It was initially called a Buckminsterfullerene, which is usually reduced to buckyball or *fullerene,* a more manageable mouthful. The buckyball is also sometimes called C60 for the 60 carbon atoms it contains.

Tony Haymet, an Australian researcher at the University of California at Berkeley, published a paper around the same time that suggested the existence of this compound, which he called footballene. He chose that name because the hexagons and pentagons formed a pattern that is the same as the one on a soccer ball. Sadly for Tony, the name didn't stick.

When the discovery of buckyballs was announced, the scientific community quickly recognized their usefulness and Curl, Kroto, and Smalley won the Nobel Prize in Physics in 1996.

The work that lead to the discovery of buckyballs was accomplished with equipment in Smalley's lab at Rice University, which led to Rice becoming one of the leading universities for nanotechnology research and education.

Studying a key proponent of the bottom-up approach

Another pioneer in the development of nanotechnology was Eric Drexler. His book, *The Engines of Creation: The Coming Era of Nanotechnology,* published in 1986, is a seminal text in the field. He was also the first person to be granted a PhD in Molecular Nanotechnology.

In 1977, while still a student at MIT, Drexler outlined the fundamental bottom-up approach to nanotechnology in which he described how we might manipulate individual atoms and thereby synthesize materials. We could use tiny machines called *molecular assemblers,* he postulated, to build just about anything using common chemicals. Drexler further suggested that by using these techniques, we could build machines called *nanorobots* that are smaller than individual organic cells. He predicted that we could infuse these cell repair machines into a person's bloodstream to cure diseases. By releasing this same type of nanorobot into our air, we might also rid ourselves of pollution. Enthusiasm for this vision of nano in the world's scientific community was off and running.

The great debate

Richard Smalley, one of the trio that discovered buckyballs and a Nobel Prize winner, had admired Drexler's ideas in earlier years. But at some point he came to think that Drexler's vision was scientifically impractical. In a long-distance debate, the two exchanged a famous series of letters. Smalley warned against a frightening future in which self-replicating nanorobots would take over the world.

The debate is often referred to as the "fat fingers problem" because Smalley doubted that we could come up with tools small enough to successfully manipulate atoms. This debate is a great example of how science is moved forward by differences of opinion and the questioning of existing theories.

Discovering nanotubes

After the discovery of buckyballs came the discovery of tiny needles of carbon that run anywhere from 1 nm to 100 nm in diameter. These tubes are made up of carbon atoms connected in hexagons and pentagons like buckyballs but in a cylindrical shape. They were discovered in 1991 by Sumino Iijima, who named them *carbon nanotubes.*

Iijima never won the Nobel Prize, but he did receive the Asahi Award, the Japan Academy Prize, and one of the most coveted prizes for scientific discovery, the Benjamin Franklin Medal in Physics.

Iijima didn't pull his discovery out of a scientific vacuum. Back in 1959 a researcher named Roger Bacon had taken pictures of carbon nanotubes, and in the 1980s another researcher named Tennant applied for a patent for a process he'd created to produce nanotubes. Smalley himself proposed a theory that nanotubes grow from buckyballs. But it wasn't until 1991 that Iijima was able to obtain high-magnification photos of carbon nanotubes and prove their existence.

Iijima examined some carbon soot under an electron microscope and realized that small cylinders of matter consisting of a lattice of carbon atoms appeared in the sample along with buckyballs. He realized that carbon nanotubes are buckyballs whose ends haven't folded around to create a sphere. Figure 2-2 shows the shape of a carbon nanotube.

Nanotubes come in two types: *single-walled nanotubes* (SWNTs) and *multi-walled nanotubes* (MWNTs). The SWNT forms a single layer cylinder and is the most commonly used type of nanotube. The MWNT is made up of multiple cylinders tucked into each other, as shown in Figure 2-3.

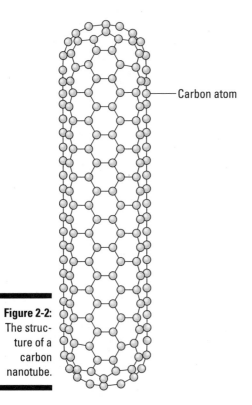

— Carbon atom

Figure 2-2:
The struc-
ture of a
carbon
nanotube.

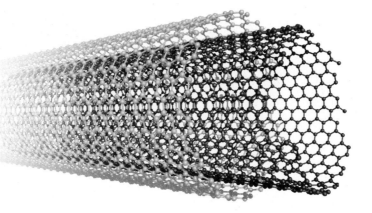

Figure 2-3:
A multi-
walled
carbon
nanotube.

Introducing the National Nanotechnology Initiative

An important step in the development of nanotechnology came in 2000 when President William Clinton introduced the National Nanotechnology Initiative, or NNI. This initiative was in part a result of the realization that a great many scientists in several organizations and in several countries were all working on pieces of the nanotechnology puzzle. The NNI was therefore created to be a coordinated program among several U.S. agencies to bring together much of that research and move the field forward.

In his announcement of NNI, Clinton also announced funding for nanoscale efforts that practically doubled previous funding amounts to a whopping $495 million starting in 2001. But the river of money didn't stop there. NNI, whose web site is shown in Figure 2-4, got great press and an enthusiastic reception; this caused more investment in the field by others, including universities, states, businesspeople, and even other governments.

NNI coordinates nanotechnology R&D (research and development) funding among approximately 25 agencies. The focus at NNI is on developing new products and techniques that could provide advances in fields such as medicine or improve the environment and building an educated workforce to support the discipline of nanotechnology. NNI also funds studies on the effect of nanotechnology on our environment and society.

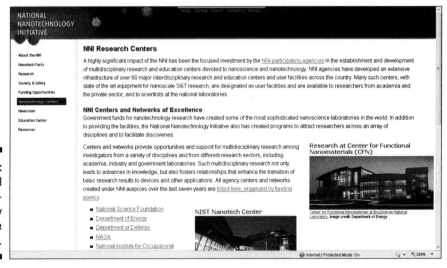

Figure 2-4:
National
Nano-
technology
Initiative
web site.

Exploring the role of Dr. Roco

Dr. Mihael Roco was an important player in the creation of the National Nanotechnology Initiative. He helped envision its structure and focus, and chaired NNI during a time when its budget grew to one billion dollars.

Roco won the 2002 Best of Small Tech Award, which called him the "Leader of the American Nanotechnology Revolution." Scientific American included him as one of 2004's top 50 technology leaders. He is generally acknowledged as the person most responsible for the level of investment and support that nanotechnology enjoys today.

To show you the scope of NNI's efforts, agencies involved include the following:

- Department of Defense
- Department of Energy
- NASA (National Aeronautics and Space Administration)
- National Institute for Occupational Safety and Health
- National Institute of Standards and Technology
- National Institutes of Health
- National Science Foundation

Eying Today's Nano Playing Field

Exactly who is involved in today's nanotechnology efforts? Many types of organizations are in relationships with each other to fund and perform research, develop and patent commercial products, and find solutions to world problems such as curing cancer and cleaning up our polluted planet. In this section, we explore this interesting mix of commercial, government, and academic folks.

Government funding for research labs

There are several interesting examples of the synergies among organizations working on nano.

Funding nanotechnology state by state

Government funding is key to the intricate network of nanotechnology research and development. In the United States, the NNI leads the way in federal funding but individual states are also involved. For example, the Pennsylvania NanoMaterials Commercialization Center supports nanomaterials research and development by Pennsylvania companies and individuals. They make grants to universities and companies focused on materials development. Several more states, such as Washington and California, also have nano funding efforts in place.

In the United States, one of the areas of focus of NNI and its associated agencies is the establishment of more than 60 nanotechnology research and education centers. These centers also provide facilities to users such as companies or researchers, sparing them the cost of purchasing high-price equipment to move along their nano-based product development.

In another example, the U.S. Department of Energy's Office of Science has created five facilities called Nanoscale Science Research Centers. These centers are located in National Labs scattered around the country: Argonne National Laboratory in Illinois; Brookhaven National Laboratory in New York State; Lawrence Berkeley National Laboratory in California; Oak Ridge National Laboratory in Oak Ridge, Tennessee; and Sandia National Laboratories in New Mexico.

The goal of these facilities is to encourage the development and characterization of new nanomaterials. Each research center has a number of focus areas that draws upon the expertise and equipment of the National Lab where they are located.

Some groups target certain aspects of nanotechnology development. The Center for Nanoscale Science and Technology (CNST) NanoFab facility in Maryland is part of the National Institute of Standards and Technology. Their mission is to solve nanoscale measurement problems that can hamper the progress of nanotechnology research.

Fighting disease

Several research programs are focused on healthcare and nanotechnology. For example, progress in the fight against cancer often requires the sharing of ideas and tools. The National Cancer Institute, in association with the National Institute of Standards and Technology and the U.S. Food and Drug Administration, has established a Nanotechnology Characterization Laboratory in Maryland. The mission of this facility is to perform preclinical efficacy and toxicity testing of nanoparticles to accelerate the transition of nanoparticles into clinical applications.

The NCI Alliance for Nanotechnology in Cancer, whose web site is shown in Figure 2-5, is another collection of organizations working to eradicate cancer using nanotechnology tools and techniques. This alliance consists of eight Centers of Cancer Nanotechnology Excellence located at various universities.

Many other healthcare-focused organizations are working on curing heart disease, diabetes, and other life-threatening conditions using nanotechnology.

See Chapter 9 for more about the application of nanotechnology to medicine.

Making a buck: The role of companies in nano development

The cold, hard fact is that to develop any technology, somebody somewhere has to find a way to monetize it. For that reason, you will find that many movers and shakers in the corporate world are interested in nanotechnology.

Many companies developing nanotechnology products have licensed techniques that were developed at universities or national labs. They take these techniques and work to develop products. They pay to put these products through the phases of development, testing, and marketing, and then share in the profits.

Figure 2-5:
The National Cancer Institute Alliance for Nanotechnology in Cancer.

Some companies do their own nanotechnology research. These companies range from ones focused on specific areas of nanotechnology, such as Zyvex, which was formed to focus on molecular nanotechnology, to large companies conducting significant nanotechnology research programs intended to improve their products. In the latter category are companies such as IBM and Hewlett Packard.

Some companies manufacture nanomaterials to sell to other companies to incorporate into their products. These nanomaterial companies range from large, established companies that build plants to supply great quantities of material, to small companies that license techniques developed at universities or national labs to produce specialized nanomaterials.

Several companies pool their resources, spreading the expense of research. For example, a group of oil companies and oil services companies has combined with Rice University and the University of Texas at Austin to form the Advanced Energy Consortium. Each company contributes the funds needed by the consortium to research nanotechniques to improve the discovery and recovery of oil.

Another example of companies being involved with universities and government in nano research is the Institute for Soldier Nanotechnology's Industry Consortium, or ISN. Located at MIT, ISN is funded by the U.S. Army, ISN has 14 partners in the private sector. For example, Raytheon and DuPont formed an industry consortium that focuses on creating products that could benefit the military, such as lighter weight packs for soldiers in the field.

Educating our workforce

Universities and funding groups are heavily into research, but another strong focus behind those ivy walls is educating our future workforce in the skills needed to pursue nanotechnology careers.

In another U.S. example, the National Science Foundation has funded 18 Nanoscale Science and Engineering Centers. In addition to research, they support education programs for K–12 and university education in nanotech-related fields.

If you're interested in education in nanotechnology, take a look at Chapter 16 for more about some specific programs offered by leading institutions.

Developing nano internationally

Although the preceding examples of nanotechnology initiatives are taken from the United States, the first country to develop such cooperative nano-focused efforts, these types of programs are by no means limited to the United States. More than 60 countries have some form of nano initiative in the works.

Indeed, China, Germany, Japan, and Korea are just a few of the countries that are funding major nanotechnology research and development to stimulate economic growth and address societal challenges such as healthcare and pollution. Several third-world countries also have their own programs because they don't want to be left out of the incredible benefits nano will provide, which they so desperately need.

But it's not just governments getting into nano around the globe. International companies such as Mitsubishi Corporation (automobiles, Japan), Fuji Film (cameras, Japan), Saudi Aramco (oil, Saudi Arabia), Carl Zeiss (optical lenses and microelectronics, Germany), LG Chem (appliances and cell phones, South Korea), and L'Oreal (cosmetics, France) are studying uses of nanotechnology in their industries.

The graph in Figure 2-6, produced by Lux Research, looks at the combination of nanotechnology research and development strength in several countries.

Although competition exists among countries to discover the next great nano application, the international cooperation in this field is amazing. For example, about 25 percent of all scholarly papers about nano are written by coauthors from different countries working together.

According to Phillip Shapira of the Georgia Institute of Technology, who coauthored a study of international nano collaboration: "Despite ten years of emphasis by governments on national nanotechnology initiatives, we find that patterns of nanotechnology research collaboration and funding transcend country boundaries. For example, we found that U.S. and Chinese researchers have developed a relatively high level of collaboration in nanotechnology research. Each country is the other's leading collaborator in nanotechnology R&D."

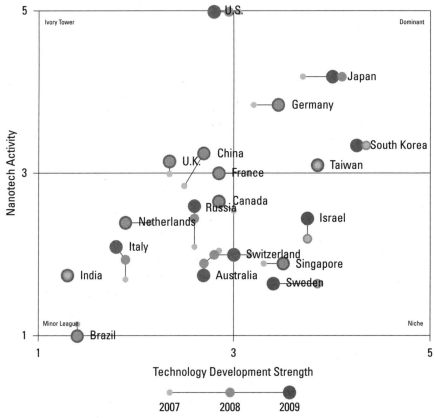

Figure 2-6: Nano-technology research and development activity by country.

Courtesy of Lux Research

Understanding Where Nano Is and Where It's Going

Much of what people talk about doing with nano lies in the future. However, you can find many examples of nanotechnology making a difference today, as well as many products and techniques under development.

In this last section, we provide an overview of which nano products and processes are a reality, which are under development, and which are only a glimmer in some nanotechnologist's eye.

Getting nano today

Nanotechnology makes it possible to achieve several benefits when you manufacture materials. For example, nanomaterials can be stronger and more lightweight than their non-nano counterparts. Nano also makes it possible to make materials smaller, a key aspect of building computer chips, for example. In addition, nanoparticles can help fibers resist stains and repel water. Used as catalysts in chemical reactions, nanoparticles can make processes more efficient and reduce the amount of energy they require. Nano also has several applications in healthcare.

Here's a list of the types of things nano is making possible today. Nano is being used

- ✔ To make strong lightweight equipment ranging from tennis racquets to windmill blades

- ✔ To clean up industrial solvents contaminating groundwater

- ✔ To protect clothing with nanoparticles that shed water or stains

- ✔ As catalysts to make chemical manufacturing more efficient while saving energy and keeping waste products to a minimum

- ✔ As a coating on countertops that kills bacteria

- ✔ In sunscreens to provide protection from UV rays without producing a thick white residue

- ✔ In wound dressings to rapidly stop bleeding in trauma patients

- ✔ As a film on glass to stop water from beading and dirt from accumulating

- ✔ In paints to prevent corrosion and the growth of mold as well as to provide insulation

- ✔ To make integrated circuits with features that can be measured in nanometers (nm), allowing companies to make computers chips that contain billions of transistors

- ✔ In bandages to kill germs

- ✔ For coatings in heavy-duty machinery, such as ships and the oil industry, to make equipment last longer

- ✔ In plastic food packaging to keep oxygen out so the food spoils at a much slower rate

Exploring efforts under development

Many uses of nanotechnology are at various stages of development, having moved from the concept stage into testing and beyond. Although you can't go shopping for these products or services yet, they may become available in only a few years to a decade or so.

Several of these developments are in the medical arena, including the following:

- Targeted drug delivery, such as delivering chemotherapy drugs directly to cancer tumors, avoiding the harm to healthy organs that currently occurs with chemotherapy

- Better imaging, with devices such as an MRI (magnetic resonance imaging), by attaching nanoparticles to diseased cells that increase the imaging signal, providing better images of tumors

- Sensors used for medical diagnosis to allow earlier detection of diseases in the patient; or to detect bacteria, such as salmonella, in foods

- The capability to regrow cartilage in joints to relieve arthritis pain

- The use of nanofiber materials that can replace or repair blood vessels and help patients with vascular or heart disease

- Burn dressings that release antibiotics when a burn becomes infected, enabling more rapid recovery

- Enabling the administration of drugs orally that were traditionally given with a shot, such as insulin

Another big area under development is nanoenergy. Here are some examples of what you might see in a few years:

- Longer-range electric cars with electrical power stored in batteries that have greater capacity than current batteries (by five or more times)

- Lower-cost solar cells made possible by using nanomaterials that reduce the energy cost used in solar cell manufacturing as well as the material and installation costs

- Strong, lightweight materials that reduce the weight of spacecraft and aircraft, thereby causing them to use less fuel

Nanotechnology is also being explored as an avenue for cleaning up our environment, including efforts to

- Clean up our water supply both by reducing the cost of producing drinking water from salt water and by removing contaminates such as metal or bacteria from polluted water

- Clean up our air by removing pollutants ranging from volatile organic compounds found in indoor air to carbon dioxide generated in power plants

These lists are by no means exhaustive, but they give you a good idea of the types of uses of nanotechnology being developed today.

For more about these three key development areas, see Chapter 9 (medicine), Chapter 10 (energy), and Chapter 11 (environment).

Eying pie in the sky

Even though much has been accomplished in developing nanotechnology applications, much more is still deemed possible. Three of the most fascinating examples — all in the speculative phase — are the space elevator, molecular manufacturing, and cellular repair.

Climbing into space

The space elevator, which would allow us to send items into space relatively cheaply, involves a cable made incredibly strong with the use of carbon nanotubes. The cable might be connected to the top of an asteroid that orbits our planet and to an anchor station somewhere in one of our oceans. We'd send equipment and materials up to space using space elevator cars. Solar cells located on the space elevator cars would use light to run the elevator, amounting in huge savings over the cost of rocket fuel to send a ship into space.

Yearly space elevator competitions are conducted by the Elevator 2010 group. These competitions invite creative people to produce prototypes for the space elevator and vie for some hefty cash prizes.

Replicating materials

If you're a *Star Trek* fan, you remember the replicator, a device that could produce anything from a space age guitar to a cup of Earl Grey tea. Your favorite characters just programmed the replicator, and whatever they wanted appeared. Researchers are working on developing a method called *molecular manufacturing* that may someday make the *Star Trek* replicator a reality. The gadget these folks envision is called a *molecular fabricator;* this device would use tiny manipulators to position atoms and molecules to build an object as complex as a desktop computer. Researchers believe that we can use raw materials to reproduce almost any inanimate object using this method.

Molecular fabricators may be available to anybody, anywhere in about 20 years or so. When fabricators are available, you'll be able to program an item's design into the machine, and that item could be produced cheaply and in large quantities. This could significantly improve living conditions in regions that do not have easy access to manufactured goods. For example, water filters could be produced to help in regions with contaminated water supplies, and solar cells could make electricity available in the remotest jungle or desert.

The existence of molecular fabricators could cause interesting ethical challenges as well as tremendous potential for upheaval in our economy. See Chapter 13 for more about ethics and regulatory concerns related to nanotechnology.

Repairing cells

Are you more interested in the Fountain of Youth than replicating solar cells? Then you'll be glad to hear that techniques for building nanorobots are being developed that should make the repair of our cells possible. As we age, DNA in our cells is damaged by radiation or chemicals in our bodies. Nanorobots would be able to repair the damaged DNA and allow our cells to function correctly.

This capability to repair DNA and other defective components in our cells could not only keep us healthy but also potentially restore our bodies to a more youthful condition.

For more about how these efforts are evolving, visit our web site at www. understandingnano.com where we provide regular updates on advances in nano.

Chapter 3

Building Blocks: Nanomaterials

*O*ur world is made up of various elements, as you'll remember if you sat and stared at the periodic table in science class for as many hours as we did. Those elements can all be broken down into nanoparticles. However, some nanoparticles are more useful than others.

In this chapter, we cover the carbon-based nanoparticles — such as buckyballs, nanotubes, and graphene — as well as nanomaterials based on a sampling of other materials such as gold and iron that have proved useful to nanoresearchers for a variety of reasons.

Carbon-Based Materials

Our world is very carbon friendly (for example, all life forms on our planet are carbon based). This could be because carbon atoms form very strong covalent bonds, bonds in which atoms share electrons with each other. In fact, the world's most popular bling, diamond, is one of the toughest materials known and is made up entirely of carbon atoms. In a diamond, each carbon atom is covalently bonded to four other carbon atoms in a three-dimensional lattice that makes it very strong indeed, as illustrated in Figure 3-1.

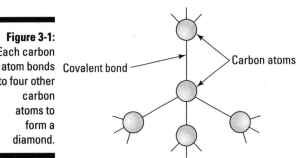

Carbon atoms are also very versatile, in that they can form covalent bonds to many other types of atoms, resulting in the formation of many other materials. Molecules that make up materials ranging from wood to the cells in our bodies are composed of carbon atoms covalently bonded with other types of atoms, which give those molecules different properties.

Buckyballs

Buckyballs, also called fullerenes, were one of the first nanoparticles discovered. This discovery happened in 1985 by a trio of researchers working out of Rice University named Richard Smalley, Harry Kroto, and Robert Curl.

Buckyballs are composed of carbon atoms linked to three other carbon atoms by covalent bonds. However, the carbon atoms are connected in the same pattern of hexagons and pentagons you find on a soccer ball, giving a buckyball the spherical structure shown in Figure 3-2.

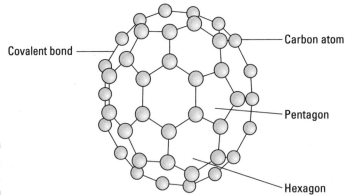

Figure 3-2:
A buckyball.

The most common buckyball contains 60 carbon atoms and is sometimes called C60. Other sizes of buckyballs range from those containing 20 carbon atoms to those containing more than 100 carbon atoms.

The covalent bonds between carbon atoms make buckyballs very strong, and the carbon atoms readily form covalent bonds with a variety of other atoms. Buckyballs are used in composites to strengthen material. Buckyballs have the interesting electrical property of being very good electron acceptors, which means they accept loose electrons from other materials. This feature is useful, for example, in increasing the efficiency of solar cells in transforming sunlight into electricity, as discussed in Chapter 10.

Carbon nanotubes

Another significant nanoparticle discovery that came to light in 1991 was carbon nanotubes. Where buckyballs are round, nanotubes are cylinders that haven't folded around to create a sphere. Carbon nanotubes are composed of carbon atoms linked in hexagonal shapes, with each carbon atom covalently bonded to three other carbon atoms. Carbon nanotubes have diameters as small as 1 nm and lengths up to several centimeters. Although, like buckyballs, carbon nanotubes are strong, they are not brittle. They can be bent, and when released, they will spring back to their original shape.

One type of carbon nanotube has a cylindrical shape with open ends, as shown in Figure 3-3.

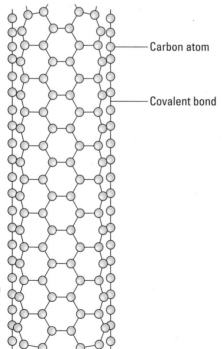

Carbon atom

Covalent bond

Figure 3-3: A carbon nanotube.

Another type of nanotube has closed ends, formed by some of the carbon atoms combining into pentagons on the end of the nanotube, as shown in Figure 3-4.

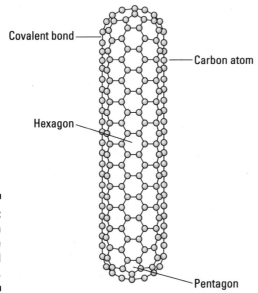

Covalent bond

Carbon atom

Hexagon

Pentagon

Figure 3-4:
A carbon
nanotube
with closed
ends.

The properties of nanotubes have caused researchers and companies to consider using them in several fields. For example, because carbon nanotubes have the highest strength-to-weight ratio of any known material, researchers at NASA are combining carbon nanotubes with other materials into composites that can be used to build lightweight spacecraft, as discussed in more detail in Chapter 12.

Carbon nanotubes can occur as multiple concentric cylinders of carbon atoms, called multi-walled carbon nanotubes (MWCTs) and shown in Figure 3-5. Logically enough, carbon nanotubes that have only one cylinder are called single-walled carbon nanotubes (SWCTs). Both MWCT and SWCT are used to strengthen composite materials. See Chapter 5 for more about nanocomposites.

Electrical properties of nanotubes

The electrical properties of carbon nanotubes depend on how the hexagons are orientated along the axis of the tube. Figure 3-6 shows the three orientations that are possible: armchair, zigzag, and chiral.

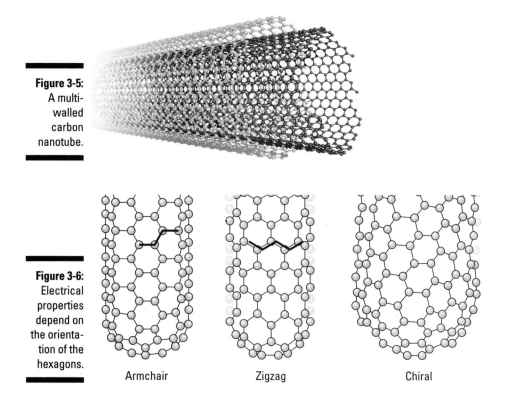

Figure 3-5:
A multi-walled carbon nanotube.

Figure 3-6:
Electrical properties depend on the orientation of the hexagons.

Armchair   Zigzag   Chiral

Carbon nanotubes with the hexagons orientated in the configuration labeled armchair (hexagons are lined up parallel to the axis of the nanotube) have electrical properties similar to metals. When you apply a voltage between two ends of an armchair nanotube, a current will flow. An armchair carbon nanotube is, in fact, a better conductor than the copper normally used in electrical wire, or any other metal.

Researchers are developing methods to spin carbon nanotubes together to make low-resistance electrical wires that could transform the electrical power grid, as we discuss in Chapter 5, as well as reduce the power consumed and the weight of wiring in such power- and weight-sensitive uses as spacecraft and airplanes. Another use that these armchair carbon nanotubes are being considered for is to connect devices in integrated circuits. As devices in integrated circuits become smaller, it may not be possible to pattern narrow enough metal lines, so researchers are considering using armchair carbon nanotubes to replace the metal lines, as discussed in Chapter 6.

The next two possible orientations of hexagons in carbon nanotubes share electrical properties similar to semiconductors. Those with the hexagons oriented in a circle around the nanotube have a configuration labeled zigzag. Those with a twist to the nanotube so the hexagons do not form any line are called chiral. These two configurations of nanotubes will only conduct an electric current when extra energy in the form of light or an electric field is applied to free electrons from the carbon atoms. Semiconducting nanotubes could be useful in building the ever smaller transistors used by the hundreds of millions in integrated circuits for all kinds of electronic devices.

Another interesting property of carbon nanotubes is that their electrical resistance changes significantly when other molecules attach themselves to their carbon atoms. Companies are using this property to develop sensors that can detect chemical vapors such as carbon monoxide or biological molecules, which we discuss in detail in Chapter 6.

Bonding nanotubes

The carbon atoms in nanotubes are great at forming covalent bonds with many other types of atoms for several reasons:

- Carbon atoms have a natural capacity to form covalent bonds with many other elements because of a property called electronegativity. Electronegativity is a measure of how strongly an atom holds onto electrons orbiting about it. The electronegativity of carbon (2.5) is about in the middle of the range of electronegativity of various substances from potassium (0.8) to fluorine (4). Because carbon has an electronegativity in the middle of the range, it can form stable covalent bonds with a large number of elements.

- All the carbon atoms in nanotubes are on the surface of the nanotube and therefore accessible to other atoms.

- The carbon atoms in nanotubes are bonded to only three other atoms, so they have the capability to bond to a fourth atom.

These factors make it relatively easy to covalently bond a variety of atoms or molecules to nanotubes, which changes the chemical properties of the nanotube. (This method is called functionalization and is discussed in more detail in Chapter 5.)

Taking this bonding thing further, if the molecules attached to the carbon nanotubes also attach to carbon fibers, the functionalized carbon nanotubes can bond to the fibers in a composite, producing a stronger material.

Graphene

Although carbon can form three-dimensional lattices by bonding with four other carbon atoms to form diamond, it can also form two-dimensional sheets (a sheet of paper has only two dimensions, for example) when it bonds to three other carbon atoms. These sheets are called *graphene*.

Researchers have only recently (2004) been successful in producing sheets of graphene for research purposes, though they all probably had a handy form of graphene in their pocket protectors. Common graphite is the material in pencil lead, and it's composed of sheets of graphene stacked together. The sheets of graphene in graphite have a space between each sheet, as illustrated in Figure 3-7, and the sheets are held together by the electrostatic force called van der Waals bonding.

Figure 3-7:
Sheets of graphene held together by van der Waals bonding make graphite.

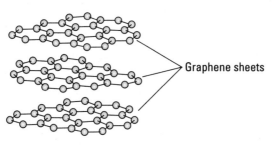

Graphene sheets

Graphene sheets are composed of carbon atoms linked in hexagonal shapes, as shown in Figure 3-8, with each carbon atom covalently bonded to three other carbon atoms. Each sheet of graphene is only one atom thick, and each graphene sheet is considered a single molecule. Graphene has the same structure of carbon atoms linked in hexagonal shapes to form carbon nanotubes, but graphene is flat rather than cylindrical.

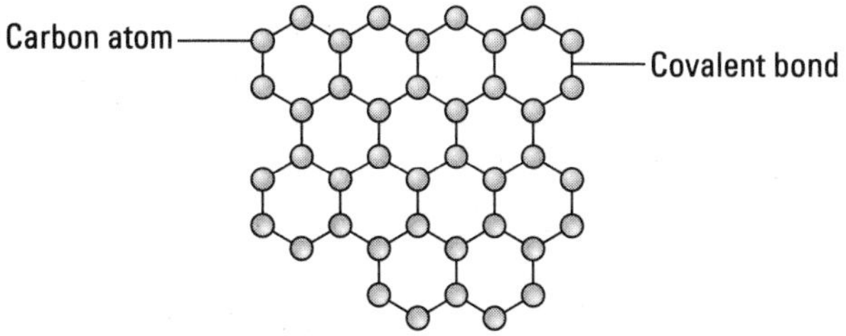

Carbon atom

Covalent bond

Figure 3-8:
A graphene sheet.

Because of the strength of covalent bonds between carbon atoms, graphene has a very high tensile strength. (Basically, tensile relates to how much you can stretch something before it breaks.)

In addition, graphene, unlike a buckyball or nanotube, has no inside because it is flat. Buckyballs and nanotubes, in which every atom is on the surface, can interact only with molecules surrounding them. For graphene, every atom is on the surface and is accessible from both sides, so there is more interaction with surrounding molecules.

Finally, in graphene, carbon atoms are bonded to only three other atoms, although they have the capability to bond to a fourth atom. This capability, combined with great tensile strength and the high surface area to volume ratio of graphene may make it very useful in composite materials, which are discussed in Chapter 5. Researchers have reported that mixing graphene in an epoxy resulted in the same amount of increased strength for the material as was found when they used ten times the weight of carbon nanotubes.

A key electrical property of graphene is its electron mobility (the speed at which electrons move within it when a voltage is applied). Graphene's electron mobility is faster than any known material and researchers are developing methods to build transistors on graphene that would be much faster than the transistors currently built on silicon wafers.

Another interesting application being developed for graphene takes advantage of the fact that the sheet is only as thick as a carbon atom. Researchers have found that they can use nanopores to quickly analyze the structure of DNA, as discussed in Chapter 9. When a DNA molecule passes through a nanopore which has a voltage applied across it, researchers can determine the structure of the DNA by changes in electrical current. Because graphene is so thin, the structure of a DNA molecule appears at a higher resolution when it passes through a nanopore cut in a graphene sheet.

Diamondoid

The name *diamondoid* may make you think of those sappy ads for engagement rings that encourage you to fork over three or four months of your hard-earned money to express your love. There's a reason why you may think that: Diamondoid is composed of carbon atoms linked to four other carbon atoms by covalent bonds in the same lattice structure that gives the diamond in an engagement ring its strength. Unlike natural diamond, diamondoid has great strength without diamond's naturally occurring flaws.

The carbon-to-carbon covalent bonds in diamondoid would result in material that has mechanical properties similar to carbon nanotubes.

You can't run out and buy diamondoid today because it's only in the very early stages of development. Scientists are hard at work on producing diamondoid structures by placing carbon atoms one at a time using a process called *mechanosynthesis.* They will eventually be able to use mechanosynthesis to build structures, such as airplane wings, that are incredibly strong. Projections are that the strength-to-weight ratio for diamondoid will be about 50 times that of high strength steel. This property makes it possible for us to use diamondoid to create materials that weigh a fraction of their current weight, which could make for strong and lighter weight products such as cars and planes.

Diamondoid may also be used in smaller structures, for example the cell-sized nanorobots that we discuss in Chapter 4.

The Runners-Up, Noncarbon Nanoparticles

Any material, such as gold, at the nanoscale is a nanoparticle. Nanoparticles are so small they contain just a few atoms to a few thousand atoms, as opposed to bulk materials that might contain billions of atoms. This difference in size causes nanomaterials to have unique characteristics. Although the versatility of carbon makes carbon-based nanomaterials especially useful to nanotechnologists, nanomaterials composed of other elements do some things better.

In this section, we introduce you to the most common materials used to construct nanoparticles. These nanoparticles commonly come in one of these forms:

- A *nanoparticle* (which may include any irregularly shaped form) is any particle with a diameter ranging from 1 to 100 nanometers.
- A *nanowire* is a wire with a diameter of less than 100 nanometers of any length.
- A *nanofilm* is a film with a thickness of less than 100 nanometers.
- A *nanotube* is a hollow cylinder with a diameter of less than 100 nanometers.
- A *nanorod* is a solid cylinder with a diameter of less than 100 nanometers.
- A *quantum dot* is a semiconductor nanoparticle.

Iron and iron oxide

Iron is an element that, in its bulk form, is used in such everyday settings as stair railings and the structural beams in cars or buildings. Iron is also present in water and in our bloodstream, where it helps to transport oxygen. Iron is one of the materials that we can use to make magnets due to the way electrons orbit each atom. And, as we all know, iron rusts when you combine iron and oxygen to form iron oxide. It turns out that nanoparticles of both iron and iron oxide can be quite useful. We start with that rusty orange substance you may have struggled to get rid of for years, iron oxide.

All electrons spin; this is a fact of life. A magnet is formed by aligning the spin of unpaired electrons to a magnetic field. Each iron atom has four unpaired electrons arranged around each atom, which means they can be aligned to turn any piece of iron into a magnet.

Iron oxide

If iron is left in the rain it will rust, and rust is composed of iron oxide, a molecule that contains three atoms of iron and four atoms of oxygen. Like iron, iron oxide has magnetic properties. Iron has four unpaired electrons, whereas iron oxide has only two unpaired electrons. Because the unpaired electrons make a material magnetic, iron oxide is less magnetic than iron. Iron oxide is therefore called a paramagnetic material. The paramagnetic properties of iron oxide nanoparticles are not changed from the bulk material except that these tiny particles can go where larger particles never could.

For example, in magnetic resonance imaging (MRI), you get a better image if paramagnetic nanoparticles are attached to the object you're taking an image of. For that reason, researchers are functionalizing iron oxide nanoparticles by coating them with molecules that are attracted to cancer tumors to provide a better MRI image. We discuss how nanotechnology is improving medical imaging in Chapter 9.

Making nanoparticles that have a core made of iron oxide nanocrystals surrounded by nanoporous silica can improve not only the MRI images of tumors but also give researchers control over the release of therapeutic drugs. We discuss how nanotechnology is improving drug delivery in Chapter 9.

A problem for tens of millions of people around the world is the presence of arsenic, a naturally occurring substance in soil that can dissolve in water, including well water. Robert Bunsen, the developer of the Bunsen burner, which most of us encountered in our high school science labs, determined in the 1830s that when you mix ferric oxide with arsenic, you get a potion that both the fluids in your body and water cannot break down. Based on Bunsen's discovery, scientists started using iron oxide in filters to remove arsenic from water. We discuss how nanotechnology is used to purify drinking water in Chapter 11.

Iron

Iron nanoparticles also retain iron's magnetic properties. What's interesting is that, like iron oxide, these magnetic nanoparticles have increased surface area. This allows the iron nanoparticles to be useful in both medical imaging and cleaning up pollutants in groundwater.

Researchers are investigating the use of iron nanoparticles as the next step beyond iron oxide nanoparticles for medical imaging (as discussed in Chapter 9) and treatments such as the following:

- **Targeted drug delivery:** Using nanoparticles with an iron core, drug delivery can be guided by a magnetic field to a particular region of a patient's body, as discussed in Chapter 9.

- **Improved MRI imaging and treatment:** After an MRI image shows that nanoparticles are concentrated at the diseased region, an oscillating magnetic electric field would be used to vibrate the nanoparticles, creating heat to kill the diseased cells, as discussed in Chapter 9.

Iron nanoparticles are also useful in cleaning up organic pollutants in groundwater because they can donate electrons to more electronegative atoms, such as chlorine atoms, present in many of the molecules that make up organic pollutants. Donating these electrons can cause the molecules to break up into harmless molecules. Because nanoparticles can remain suspended in groundwater for a long time and are transported throughout the system, they are used to treat large areas of groundwater, as discussed in Chapter 11.

Platinum

Platinum is a rare element and therefore in demand for use in both jewelry and as a catalyst. In bulk form, platinum is one of the most effective but expensive catalysts available. You probably use platinum as a catalyst every day in the catalytic converter in your car. The platinum in a catalytic converter helps change air-polluting molecules from your car exhaust into less harmful molecules.

The atoms in molecules, such as hydrogen, bond with platinum atoms, and then the platinum atoms release the hydrogen atoms, allowing them to react with other molecules. By breaking up molecules into atoms, platinum facilitates chemical reactions and allows them to occur at a lower temperature than they could without a catalyst. Using nanoparticles of platinum increases the surface area available for a reaction and also increases the percentage of platinum atoms available for contact with molecules involved in the reaction. This difference allows researchers and manufacturers to use smaller quantities of platinum, which is important given its high cost.

These improved catalysts have a better capability to break down air pollutants (as we discuss in Chapter 11) and also reduce the cost of catalysts used in fuel cells (as we discuss in Chapter 10). But platinum comes at a high price, so nanotechnologists may choose other options in some cases.

Gold

Gold is an element used in jewelry, coins, dentistry, and electronic devices. Gold is even used in some medicines. Bulk gold is considered an inert material in that it doesn't corrode or tarnish (which is why you paid so much for that engagement ring). As with all metals, gold has good electrical and thermal conductivity. Gold's capability to resist corrosion as well as its high electrical conductivity make it useful for forming contacts in electronic devices.

Gold has been used in various medical treatments over the centuries without harmful effects. It was therefore natural for researchers to look to gold nanoparticles for medical applications rather than using elements such as platinum, which can be toxic in certain circumstances. Forming gold into nanoparticles allows researchers to use gold in areas that are too small for bulk gold to reach and brings with it new capabilities.

For targeted drug delivery uses (discussed later), it will be interesting to see whether gold nanoparticles show any benefit versus cheaper types of nanoparticles, such as iron nanoparticles. For other uses, gold nanoparticles have some clear advantages.

When gold nanoparticles get really small, with a diameter of 5 nm or less, they can be used as a catalyst to help reactions that, for example, transform air pollutants into harmless molecules, as we discuss in Chapter 11.

Using gold to clean up the air is somewhat surprising given that bulk gold is considered to be an inert material in that it doesn't corrode or tarnish. Normally, gold would be a silly material to use as a catalyst for chemical reactions because it doesn't do much. However, if you break down gold to nanosize (approximately 5 nanometers), it can act as a catalyst that can do things such as oxidizing carbon monoxide.

Researchers attach molecules to gold nanoparticles that are attracted to diseased regions of the body, such as cancer tumors, and other molecules such as therapeutic drug molecules. This enables the functionalized gold nanoparticles to be used in targeted drug delivery, which is discussed in Chapter 9.

Another property that gold nanoparticles have is the capability to convert certain wavelengths of light into heat. As with all metals, gold contains electrons that are not tied to a particular atom but free to move throughout the metal. These electrons help to conduct a current when a voltage is applied across the conductor. Depending on the size and shape of the nanoparticles, these free electrons will absorb the energy from a particular wavelength of light, at the right wavelength to make the cloud of free electrons on the surface of the gold nanoparticle resonate. It turns out that two types of gold nanoparticle shapes are more efficient in converting light into heat:

- ✔ **Gold nanorods:** These solid cylinders of gold have a diameter as small as 10 nm. By using nanorods with different combinations of diameter and length, researchers can change the wavelength of light that the nanorod absorbs.

- ✔ **Nanospheres consist of a gold coating over a silica core:** By using nanospheres with variations in the thickness of the gold coating and the diameter of the silica core, researchers can change the wavelength of the light that the nanosphere absorbs.

Various researchers are using either nanorods or nanospheres to develop methods for localized heat treatment of diseased regions of the body. This method is called hyperthermia therapy and is discussed in Chapter 9.

Silica

Silica, or silicon dioxide, is the same material used to make glass. In nature, silica makes up quartz and the sand you walk on at the beach. Unlike metals such as gold and iron, silica is a poor conductor of both electrons and heat. Despite these limitations, silica (silicon oxide) nanoparticles form the framework of silica aerogels. *Silica aerogels* are composed of silica nanoparticles interspersed with nanopores filled with air. As a result, this substance is mostly made up of air. Because air has very low thermal conductivity and silica has low thermal conductivity, they are great materials to use in insulators. These properties make nano aerogels one of the best thermal insulators known to man. We discuss silica aerogels for use in spacecraft in Chapter 12, for use in insulating your house in Chapter 7, and for use in clothing in Chapter 8.

You can also functionalize silica nanoparticles by bonding molecules to a nanoparticle that also is able to bond to another surface, such as a cotton fiber. The functionalized silica nanoparticles attach to the cotton fiber and form a rough surface that is hydrophobic (water repellent), giving an effect similar to the water repellency of lotus leaves. We discuss the use of nanoparticles to produce water-repellant fabrics in Chapter 8.

Another type of silica nanoparticle is riddled with nanoscale pores. Researchers are developing drug delivery methods where therapeutic molecules stored inside the pores are slowly released in a diseased region of the body, such as near a cancer tumor, as shown in Figure 3-9. We discuss drug delivery methods in Chapter 9.

Courtesy of Pacific Northwest National Laboratory

Figure 3-9:
Y-shaped
antibod-
ies being
carried in
nanopores
for delivery
to cancer
tumors.

Silicon dioxide nanofilms, a layer of silicon dioxide molecules that can be as thin as 1 nm, are used to provide electrical insulation between two parts of a device, such as a transistor. This method is used in making computer chips, which we discuss in Chapter 6.

Silver

Silver, like gold, is an element that is used in jewelry, coins, dentistry, and electronic devices. Silver has been used also to kill bacteria, for example, by preventing infection in wounds before antibiotics were developed. As with all metals, silver has good electrical and thermal conductivity.

Silver nanoparticles have a large percentage of atoms at the surface and are useful for killing bacteria. Nanoparticles of silver have more silver ions on the surface and are therefore being used as an antimicrobial agent in several ways, including:

✔ As an antimicrobial agent for the treatment of wounds in crystalline form

✔ Embedded in plastic food storage bins to kill bacteria from any food previously stored in the bins

✔ Destroying odors caused by microorganisms in the form of microbes that can't live around silver ions

✔ To make a low-cost filter from carbon nanotubes and silver nanowires to kill bacteria in drinking water

Titanium dioxide nanoparticles

Titanium dioxide is a molecule composed of one atom of titanium and two atoms of oxygen. Titanium dioxide absorbs ultraviolet light; this property makes titanium dioxide useful in sunscreens.

Titanium dioxide nanoparticles are photocatalysts, which means that they have the capability to use energy in light to catalyze reactions with other molecules at reduced temperatures. Although other photocatalytic materials are available, researchers have found that titanium dioxide provides the best performance in sunlight.

Another property of titanium dioxide is that it reflects all colors in the visible light spectrum, therefore the light reflected from titanium dioxide is white. This characteristic makes it useful as a white pigment in paints and may make for white residue on your skin when you slather on sunscreen.

Forming titanium dioxide nanoparticles allow researchers to form photocatalysts that are more effective because they have more surface area available to react with other molecules. Also, in the nanoparticle form, titanium dioxide can be used in creams and coatings that absorb UV without causing a white coating.

Note that titanium dioxide (a molecule composed of one atom of titanium and two atoms of oxygen) nanotubes are similar to carbon nanotubes in that they are hollow cylinders. But titanium dioxide nanotubes, instead of being composed of carbon atoms bonded together, are made up of titanium dioxide molecules bonding together to form the surface of a cylinder. The oxygen atoms bond to another titanium atom, and you end up with a cylindrical lattice in which each oxygen atom is bonded to two titanium atoms. This arrangement of oxygen atoms between each titanium atom results in a much more complicated nanotube structure.

Some companies use titanium oxide nanoparticles as part of a film that uses the energy in light to start the chemical reaction that kills bacteria on surfaces. We discuss these types of antibacterial coatings in Chapter 7.

At Penn State, a team of researchers led by Craig Grimes has come up with an ingenious method of turning captured CO_2 into methane. They use clusters of titanium dioxide nanotubes coated with a catalyst that helps convert carbon dioxide and water into methane using sunlight as the power source. We discuss this method in Chapter 11.

Researchers are also developing methods to use the photocatalytic properties of titanium dioxide nanoparticles to destroy cancer tumors. They are delivering titanium dioxide nanoparticles to cancer tumors using targeted drug delivery methods discussed in Chapter 9, then shining light on the tumor. The titanium dioxide nanoparticles use the energy from the light to add an electron to oxygen molecules, which proceed to destroy cancer cells.

Silicon

Besides being a snappy label for areas where computer technology thrives, silicon is an element whose atoms covalently bond together to form a semi-conducting material. Wafers made of silicon crystal are used as the substrate on which computer chips are built, as well as being used in many types of solar cells. When silicon is bonded to other elements, it can also form useful materials such as glass (silicon bonded with two oxygen atoms).

Forming silicon nanoparticles or nanowires can change the optical and mechanical properties of the material and make it possible to build ever smaller computer chips year after year.

Silicon nanowires can be useful in building transistors, which are used in every integrated circuit. They would function as the channel of the transistor, as discussed in Chapter 6. The channel of a transistor is the smallest feature in an integrated circuit. As transistors used in integrated circuits continue to get smaller to fit more transistors on a computer chip, silicon nanowires may allow transistors to shrink further than current methods would allow.

Researchers have grown silicon nanowires on a stainless steel substrate. Batteries built with anodes using these silicon nanowires have up to ten times the power density of conventional lithium-ion batteries. Bulk silicon cracks due to both the swelling of silicon as it absorbs lithium ions when a battery is recharged and the contraction of silicon as the battery is discharged (when the lithium ions leave the silicon). Silicon nanowires eliminate cracking.

Researchers are developing silicon nanoparticles to be used for fluorescent imaging of diseased tissue in the body, such as cancer tumors, as discussed in Chapter 9. Because silicon is a semiconductor, silicon nanoparticles are part of the class of nanoparticles called quantum dots.

Exploring the potential of quantum dots

Quantum dots may be able to increase the efficiency of solar cells. In normal solar cells, a photon of light generates one electron. Experiments with both silicon quantum dots and lead sulfide quantum dots can generate two electrons for a single photon of light. Therefore, using quantum dots in solar cells could significantly increase their efficiency in producing electric power.

Researchers are also working on the use of quantum dots in displays for applications ranging from your cell phone to large screen televisions that would consume less power than current displays. By placing different size quantum dots in each pixel of a display screen, the red, green, and blue colors used to generate the full spectrum of colors would be available. We discuss nanotechnology improvements to displays in Chapter 6.

Quantum dots are semiconductor nanoparticles that glow a particular color after being illuminated by light. The color they glow depends on the size of the nanoparticle. When the quantum dots are illuminated by UV light, some of the electrons receive enough energy to break free from the atoms. This capability allows them to move around the nanoparticle, creating a conduction band in which electrons are free to move through a material and conduct electricity. When these electrons drop back into the outer orbit around the atom (the valence band), as illustrated in Figure 3-10, they emit light. The color of that light depends on the energy difference between the conduction band and the valence band.

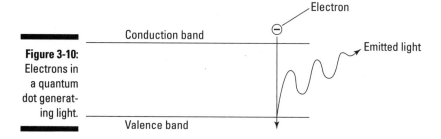

Figure 3-10:
Electrons in a quantum dot generating light.

The smaller the nanoparticle, the higher the energy difference between the valence band and conduction band, which results in a deeper blue color. For a larger nanoparticle, the energy difference between the valence band and the conduction band is lower, which shifts the glow toward red.

Many semiconductor substances can be used as quantum dots, such as cadmium selenide, cadmium sulfide, or indium arsenide. Nanoparticles of these, or any other semiconductor substance, have the properties of a quantum dot. The gap between the valence band and the conduction band, which is present for all semiconductor materials, causes quantum dots to fluoresce.

Palladium

If you're British, playing at the Palladium has one meaning, but for a chemist, palladium is an element that can be used as an option to platinum in jewelry and as a catalyst. Palladium has similar properties to platinum but costs about half as much.

In bulk form, palladium is an effective catalyst, though not as effective as platinum. When used as a catalyst, palladium allows the atoms in molecules, such as hydrogen (the hydrogen molecule contains two hydrogen atoms), to bond with its palladium atoms and then releases the hydrogen atoms, allowing them to react with other molecules.

By breaking up molecules into atoms, palladium facilitates chemical reactions and allows them to occur at a lower temperature than they would without a catalyst. For example, car manufacturers use palladium as a catalyst in the catalytic converter in your car to make air-polluting molecules from your car exhaust less harmful. Using nanoparticles of palladium increases the surface area and percentage of platinum atoms available for contact with molecules involved in the reaction. This lets you get away with smaller quantities of palladium.

Palladium nanoparticles have been shown to be effective catalysts in a variety of chemical reactions. For example, palladium nanoparticles can reduce the amount of platinum that has to be used as a catalyst in fuel cells. Palladium costs roughly half of what platinum costs, which could help reduce the cost of fuel cells.

Researchers have also found that using palladium nanoparticles can improve the performance of a catalyst. In laboratory tests, they found that a fuel cell using a catalyst made with these nanoparticles generated 12 times more current than one containing a catalyst using pure platinum. The fuel cell also lasted ten times longer. We discuss how nanotechnology is being used to improve fuel cell efficiency in Chapter 10.

Palladium can also improve the performance of iron nanoparticles used to clean up organic pollutants in groundwater. Adding palladium to iron nanoparticles allows them to contribute more electrons to the reaction, increasing the rate at which chlorinated hydrocarbons such as TCE are degraded.

Neodymium

Neodymium is an element used to make strong magnets. Neodymium's magnetic capability is due to the way electrons orbit each atom. A magnet is formed by aligning the spin of unpaired electrons to a magnetic field. Each neodymium atom has seven unpaired electrons floating around each atom that can be aligned to turn a piece of neodymium into a magnet. When neodymium is used in an alloy along with iron and boron, the combination results in small, powerful magnets that are useful in all kinds of electronic devices. The capability to make such magnets is one of the reasons that electronic devices such as laptops can be so small and lightweight.

Neodymium is a rare earth element; others include samarium, dysprosium, and praseodymium. These elements are called rare earths simply because, in their natural state, they are mixed with other elements. This mixture means that they have to be mined and extracted, which brings with it the jeopardy of environmental pollution through the mining process. Because of these complications, there is concern that we could have supply shortages of such elements, which could have a serious effect on industrial production.

Researchers are developing magnets that use less neodymium by using neodymium nanoparticles. This research is in the early stages, but the idea is that by using nanoparticles of neodymium and other materials such as iron, the coupling between the different atoms will be stronger, resulting in a stronger magnetic field even though less material is used.

Boron

Boron is an element that is used to control the rate of reactions in nuclear reactors by absorbing neutrons. Neutrons are uncharged particles that, along with protons (positively charged particles), make up the nucleus of atoms. When neutrons collide with the nucleus of atoms such as uranium, they cause the uranium atom to fission (split into two other, smaller atoms) and generate energy. Because it can absorb neutrons, boron can be used to stop that reaction.

Two types of boron atoms (called isotopes) occur naturally. The type of boron atom that occurs most frequently, about 80 percent of the time, has 5 protons and 6 neutrons in its nucleus. Because that adds up to 11 neutrons, it is called boron-11. The boron that naturally occurs the other 20 percent of the time has 5 protons and 5 neutrons and is called boron-10. It's this isotope that is good at absorbing neutrons. Nuclear theoreticians say that boron-11 is more stable than boron-10 because the neutrons are paired in boron-11 but the fifth neutron in boron-10 is unpaired. In that situation, if a neutron wanders by the unpaired neutron, it grabs hold of it.

Researchers are delivering boron nanoparticles to cancer tumors using targeted drug delivery methods discussed in Chapter 9, and then irradiating the tumor with neutrons. When boron absorbs a neutron, it splits up and sends out charged particles, called alpha particles, that destroy the cancer cells. You can't directly irradiate the tumor with alpha particles because they are short-range particles that wouldn't be able to penetrate through the body to the tumor. Instead, researchers use the longer range particle, the neutron, which is transformed into the short-range, very destructive alpha particle.

When boron absorbs neutrons it produces destructive alpha particles in a tumor, effectively concentrating the effect of the neutrons and allowing a low dose of neutron to kill the tumor. This effect is important because, although neutrons just pass through most materials, some other atoms in the body absorb neutrons that cause damage to healthy tissue. By keeping the dose of neutrons as low as possible, the damage to healthy tissue is minimized.

Boron Nitride

Boron-nitride nanotubes were discovered in 1995. Research into applications for carbon nanotubes is much further along than for boron-nitride nanotubes; but researchers are working on taking advantage of the benefits that boron-nitrite nanotubes offer.

Boron-nitride nanotubes are similar to carbon nanotubes, in that they are hollow cylinders formed by atoms connected together in hexagonal shapes. However. boron-nitride nanotubes, instead of being composed of carbon atoms, are composed of boron atoms covalently bonded to nitrogen atoms to form hexagons, as illustrated in Figure 3-11.

What's interesting is that boron-nitride nanotubes have more consistent electrical properties than carbon nanotubes. Unlike carbon nanotubes, only some of which have the electrical properties of semiconductors, all boron-nitride nanotubes have those properties. Therefore, using boron-nitride nanotubes as the transistor channel (the portion of a transistor that is opened or closed to turn a transistor on or off) in place of carbon nanotubes ensures that you have a nanotube with semiconductor properties. For more on transistors, see Chapter 6.

Researchers are also looking at the possibility of using boron-nitride nanotubes, which are almost as strong as carbon nanotubes, in composites to create strong lightweight materials, as discussed in Chapter 5. Such composites may be particularly useful for spacecraft because in a hull built from a composite containing boron-nitride nanotubes, the boron atoms can absorb neutrons from the solar wind and protect the crew and electronics.

Slightly different types of boron atoms are called isotopes. If you ever find yourself in a position to use them, be sure to specify the right kind of isotope for absorbing neutrons: boron-10, also referred to as 10B.

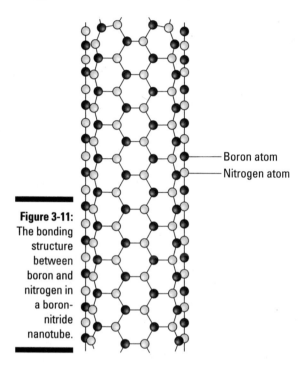

— Boron atom
— Nitrogen atom

Figure 3-11: The bonding structure between boron and nitrogen in a boron-nitride nanotube.

Researchers are constantly developing new ways to use various types of nanoparticles. For the latest updates, visit Understanding Nano's Nanoparticles web page at `www.understandingnano.com/nanoparticles.html`.

Searching for Nanoparticles

Now that you're educated about many kinds of nanoparticles, you might be wondering where you might buy this stuff. You can't find nanoparticles on aisle 3 in Walmart, but they are manufactured for various uses by a variety of companies.

Some large companies are investing significant amounts of money in building their own plants to manufacture carbon nanotubes to sell to others.

These large-scale operations should bring the price of carbon nanotubes low enough that the materials are more practical for use in more applications. Other nanoparticle manufacturers started with techniques developed in various universities that work on a smaller scale. Regardless of their size, many nanoparticle manufacturers have techniques to functionalize nanoparticles and work with their customers to integrate their nanoparticles into composite materials. Figure 3-12 shows the web site of one such producer of nanomaterials, Nanophase.

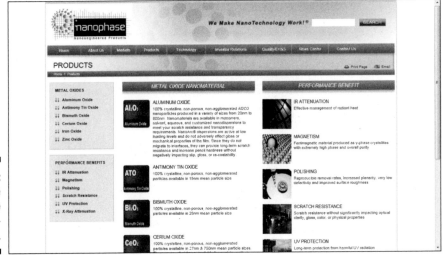

Figure 3-12: One nanoparticle manufacturer.

If you're shopping for nanomaterials, check out Nanowerk at www.nanowerk.com/phpscripts/n_dbsearch.php. They have a useful listing of suppliers listed by type of nanomaterial.

Chapter 4

Nano Tools

· ·

In This Chapter

▶ Using microscopes to view nano-sized particles

▶ Manipulating nanomaterials

▶ Building nanorobots

▶ Using nanoparticles to etch patterns on computer chips

· ·

*I*n Chapter 3, we walk you through the discovery of various particles at the nano level, such as buckyballs and nanotubes. But what's the point of discovering nanoparticles if you have no way to work with them?

That's where nano tools come in. We have several options for observing and manipulating nanoscale elements. From working with subatomic-level particles to build new materials, to working with nano-sized materials to pack more on a computer chip, these tools put nanotechnology to work in a practical way.

In this chapter, we cover specialized microscopes that help you view nanoparticles, tools to help you move atoms around, nanorobots under development that could help with tasks such as keeping your body healthy, and nanolithography for creating ever tinier integrated circuits.

Viewing Things at the Nano Level

In the tiny world of nano, the first need for researchers is to see what they're working with. A few types of microscopes being used in nano labs today do just that. They come with handy acronyms such as TEM and SEM and would stretch most of our budgets, costing anywhere from one hundred thousand to several million dollars.

Electron microscopy

Remember pulling a microscope out of a cabinet in your high school biology class and carrying it over to your lab bench? After plugging in the cord and turning on the light on the back of the microscope, you got a close up look at the microscopic creatures in pond water. For looking at nano-sized objects, the microscopes are often bigger than you, and instead of towering over the equipment as you stand at a lab bench, you sit in front of these desk-sized microscopes and they tower over you, as shown in Figure 4-1.

Figure 4-1:
An electron microscope used in working with nano-technology.

Courtesy of National Institute of Standards and Technology

The optical microscope in your biology class uses light to illuminate micro-scopic objects. An *electron microscope* uses electrons to visualize nano objects. The lamp in an optical microscope is small (because all you need to do to gen-erate light is run electricity through a filament). The electron gun in an electron microscope, however, is large because you have to accelerate the electrons before you illuminate the sample with a beam of electrons, and all that running around requires some space. An electron microscope also uses an electric field of several thousand volts to shoot electrons at a sample. The higher voltage accelerates the electrons, resolving the smaller features in the sample.

Electron microscopes come in a few varieties, the most common of which is a *scanning electron microscope* (SEM), illustrated in Figure 4-2. In an SEM, elec-trons are accelerated in the electron gun and run through a scanning coil that applies an electric field to scan the beam of electrons over the sample. (Note that all this happens in a vacuum because the electrons don't get far moving through air.) These electrons excite other electrons in the sample, which are picked up by the detector. A computer turns the signal from the detector into a picture on the monitor screen.

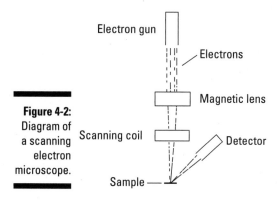

Figure 4-2:
Diagram of
a scanning
electron
microscope.

SEMs are found in labs that investigate nanomaterials. The nice thing about SEMs is that it's easy to prepare a sample. If a sample is metallic, you just put the sample in; if it's not metallic, you just coat it with gold.

How do you coat a sample with gold? It's pretty simple. Put your sample in a small sputtering chamber that contains a gold target. Argon gas flows into the chamber and is ionized by an electric field. The argon ions, in a process called sputtering (hence the name of the chamber), strike the gold target and knock some gold atoms off into the chamber. Those atoms coat your sample.

Some SEMs are capable of resolving features as small as a few nanometers. SEMs have very good depth of field, meaning that they can produce sharp images of rough samples.

If you're dealing in a really tiny realm and need to get details of a sample down to a few tenths of a nanometer (the size of an atom) to see the internal structure of the sample, you can use a transmission electron microscope (TEM). The term *transmission* in its name means that the electrons pass through the sample, as illustrated in Figure 4-3.

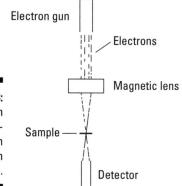

Figure 4-3:
Diagram
of a trans-
mission
electron
microscope.

Because the image is formed by the electrons passing through the sample, the sample has to be very thin. To make such a sample, you must slice and polish it; therefore, sample preparation is typically much more involved for a TEM than an SEM.

To produce images with a resolution of a few tenths of a nanometer, a TEM uses a much higher voltage to accelerate electrons than an SEM, which makes the sticker price and installation cost of a TEM much higher. Given the higher equipment cost, more difficult sample prep, and more extensive training required to use TEMs, many more SEMs are in use than TEMs. Think of SEMs as the workhorses of nanolabs and TEMs as thoroughbreds. Only those who truly need the capability to view things at an extremely small scale have a TEM in their lab.

Atomic force microscopy (AFM)

If you need an image of the surface of a sample at atomic resolution and don't need to look at the sample's internal structure, you can use a handy instrument called the *atomic force microscope* (AFM). The AFM contains a tip attached to a cantilever, similar to the needle used in record players. (You remember records, right?) The tip in an AFM is tiny, less than a nanometer in width. You aim a laser at the cantilever, as shown in Figure 4-4, and as the AFM tip moves up and down as it scans across the surface of the sample, the position of the laser beam on the detector shifts.

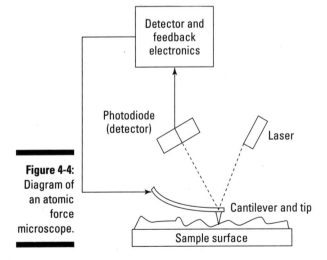

Figure 4-4: Diagram of an atomic force microscope.

This method generates a topographical image of the surface with sufficient resolution to determine the position of individual atoms, as illustrated in Figure 4-5.

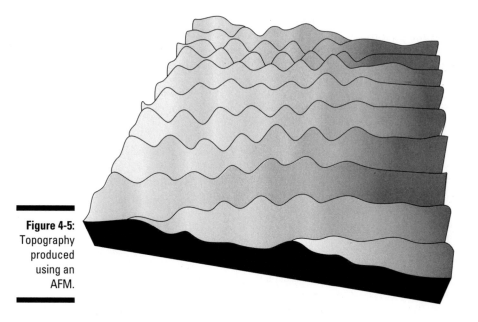

Figure 4-5:
Topography
produced
using an
AFM.

As with the SEM, an AFM is easy to use and therefore has become ubiquitous
in labs that work with nanomaterials. The AFM has higher resolution than
the SEM, but the SEM can look at a much larger area of a sample, so the two
types of instruments complement each other nicely.

Tracking molecular fingerprints with spectroscopy

When researchers build a new nanomaterial, they have this habit of want-
ing to verify its structure. Microscopes such as AFMs can give them visual
information about the structure of the new material, but other techniques
can easily provide additional information, such as what types of atoms are
bonded to each other in the material and how many electrons are shared
among these atoms.

It turns out that molecules, and portions of molecules (the pairs of atoms that
bond together to form molecules), have fingerprints, just as we do. However,
their fingerprints consist not of little swirls in the skin but of the wavelength of
the light that atoms and molecules absorb. The wavelengths of light absorbed
by various atoms and molecules produce a spectrum. Generating a spectrum
to determine the structure of molecules is called spectroscopy.

For example, the simplified spectrum labeled A in Figure 4-6 shows that carbon nanotubes absorb infrared light at a wavelength of 6452 nm and 8403 nm. The spectrum labeled B has all the absorption points shown in spectrum A, plus a peak at 5838 nm, which indicates the bonding of oxygen to some of the carbon atoms in the nanotube. Therefore, if researchers take an infrared spectrum of an unknown material and see the peaks shown in spectrum B, they know that the material consists of carbon nanotubes functionalized by adding oxygen atoms.

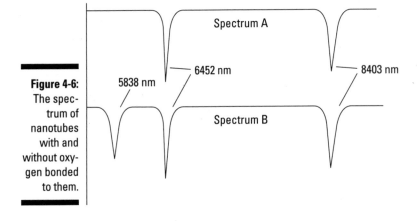

Figure 4-6: The spectrum of nanotubes with and without oxygen bonded to them.

Manipulating Atoms and Molecules

To achieve the bottom-up vision of nanotechnology (the capability to build materials by manipulating atoms, as discussed in Chapter 1) involves moving atoms and molecules to precise locations. In this section, we explore just how that's accomplished.

Moving atoms with the scanning tunneling microscope

As with the AFM, the STM (*scanning tunneling microscope*) can produce an image of the surface of a sample at atomic resolution, though it can be used only with conductive samples. Where the STM outdoes the AFM is in its capability to also move atoms around on a surface.

As shown in Figure 4-7, an STM tip narrows to a sharp point, ideally made up of just a single atom. When the tip is brought very close to the sample surface, with only about a 1 nm gap, an electrical current (called the tunneling current) occurs between the STM tip and the sample. The amount of tunneling current increases as the gap between the tip and the surface decreases. This change in the tunneling current generates a topographical image of the surface. If the STM tip, as it scans across the sample's surface, encounters an atom sitting on the surface, the gap shrinks and the tunneling current goes up.

Because the tip and the sample have no physical contact, the electrons have to tunnel across the gap between the tip and the sample to produce an electric current. The rules of quantum mechanics, which govern the behavior of subatomic particles, apply when working at this small scale, which is why this movement of electrons across a gap is called *quantum mechanical tunneling.*

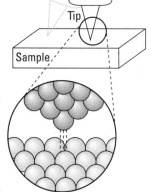

Figure 4-7:
The tip of a scanning tunneling microscope.

So exactly how does an STM move atoms? A physicist named Johannes van der Waals discovered one of the weaker forces that acts on molecules and atoms. This van der Waals force allows the STM to move atoms around. To move a particular atom to a different point on the surface of a sample, you position the STM tip above the atom. You then lower the tip to the point where the van der Waals force is strong enough to make the atom stick to another atom at the end of the STM tip when it's moved latterly. (The tip doesn't have a strong enough force to actually pick up the atom.) After the STM moves the atom to the desired spot, you raise the STM tip and the atom stays in place. Some industrious soul at the National Institute of Standards and Technology (NIST) tried this with cobalt atoms on a copper surface, moving the cobalt atoms to form the NIST logo, as shown in Figure 4-8.

Image courtesy of the National Institute of Standards

Figure 4-8:
Cobalt atoms arranged with a scanning tunneling microscope.

Molecular assemblers

When researchers can move atoms around, the potential for manufacturing at the nano level gets really interesting. Have you ever wanted to create your own iPhone, atom by atom? Well, after researchers demonstrated the ability to move atoms, the next logical step was towards creating the molecular assemblers used in atomically precise assembly, discussed in Chapters 1 and 2. In this process, you adapt STMs or AFMs to add atoms or molecules to a surface, which allows you to construct a structure atom by atom or molecule by molecule. For example, Figure 4-9 illustrates a process in which two carbon atoms are deposited on a diamond surface.

One of the key issues in designing a tip used in molecular assembly is to get the chemistry right so that the atoms you want to deposit on a target bond to the tip initially and then release from the tip at the target location and bond to target atoms. Robert Freitas, a researcher in molecular manufacturing and nanomedicine, suggests that the atoms on a tip that are bonded to carbon atoms be made of a substance such as germanium or tin. These substances form relatively weak covalent bonds with carbon. When the tip delivers the carbon atoms to the surface, the atoms form covalent bonds with nearby carbon atoms on the surface (in Freitas' research, he uses a diamond surface). This carbon-to-carbon covalent bond is much stronger than the covalent bond to atoms on the tip. As a result, when you pull away the tip, the carbon atoms stay in place on the surface.

Positionally-controlled carbon deposition using the DCB6-X dimer placement tool

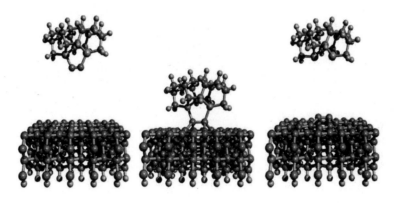

Figure 4-9:
Depositing carbon atoms in a molecular assembly process.

Researchers are planning to build molecular assemblers that can use techniques such as the one just discussed to build objects atom by atom or molecule by molecule (see Chapter 5). Freitas received a patent in 2010 covering methods to build the type of tool tip that would be used in molecular assemblers, but it will probably be a while before this kind of assembly becomes a reality.

Nanorobots

Remember Robbie the Robot from *Lost in Space?* If you think robots are cool, you'll be interested to hear that researchers are developing cell-sized robots that have a propulsion system, sensors, manipulators, and possibly even an on-board computer that can perform tasks on nanoscale objects. Nanorobots are definitely not ready for prime time, but Figure 4-10 gives a close-up of one researcher's vision of a 12-arm nanorobot, and Figure 4-11 shows an artist's conception of nanorobots at work zapping pathogens in the bloodstream. (They probably won't shout "Danger, Will Robinson!")

Figure 4-10:
A 12-arm
nanorobot
which, in
swarms,
could join
to build
virtually
anything.

Courtesy of J. Storrs Hall

Researchers are exploring the following two avenues for constructing nanorobots:

✔ One approach is to build the nanorobots from atoms or molecules using mechanosynthesis. In this method, the bulk of the nanorobot might be made up of carbon atoms, giving the nanorobot a strong, lightweight, diamondoid structure.

✔ The other method is to build nanorobots using existing biological components such as proteins. The proteins in our bodies already have capabilities, such as the capability to transform energy from molecules in the body into motion. This capability could be used to transport a nanorobot through the bloodstream. Other proteins have the capability to perform surgery at the cellular level, for example, to cut away defective portions of a DNA strand, triggering the reconstruction or repair of the strand. Researchers are working on combining biological molecules, such as proteins, either with other biological molecules or nanoparticles such as carbon nanotubes to create *bionanorobots*.

Figure 4-11:
Nanorobots
at work in
the blood-
stream.

Although it may be possible to build more capabilities into nanorobots constructed through mechanosynthesis than biologically based nanorobots, the biological versions will probably become a reality much sooner.

See more details about the design of nanorobots on our Nanorobots in Medicine page at www.understandingnano.com/nanomedicine.html.

Creating Nano-Sized Features on Computer Chips

The semiconductor industry has been in the nanotechnology business for years. They use tools and processes to etch nano-sized patterns on silicon wafers coated with a material called photoresist. Those patterns make up the circuits on the chip that allow your computer to process data. The process used to make these patterns is called *nanolithography,* which we discuss in this section.

Printing patterns with nanolithography

The integrated circuits that are the brains of your computer include nano-sized structures. For example, the smallest structure (called the minimum feature size) on the microprocessors Intel is currently shipping is 32 nanometers in width.

To create nano-sized features for integrated circuits on silicon wafers (the top-down technique described in Chapter 1) requires a machine called a stepper, which uses a technique called lithography to print a pattern on the chip. Microprocessors with a 32-nanometer feature size made with a nano-lithography process have as many as 995 million transistors packed on one computer chip. You can see a diagram of the optical portions of a stepper in Figure 4-12.

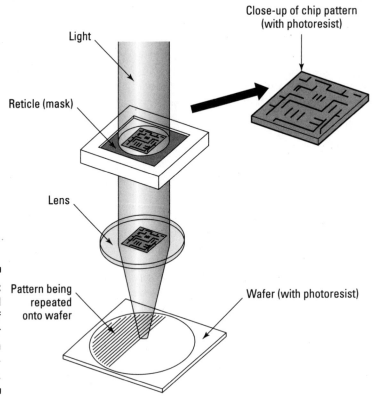

Close-up of chip pattern (with photoresist)

Light

Reticle (mask)

Lens

Figure 4-12: The optical portions of a stepper used in nanolithography.

Pattern being repeated onto wafer

Wafer (with photoresist)

In a stepper, light shines through a reticle, or photomask, which contains the pattern to be printed, and a lens focuses the pattern on photoresist coating the surface of a semiconductor wafer. The wafer is then shifted, or stepped, so that an unexposed region of photoresist moves under the optical system, which then exposes that region using UV light. This stepping continues until the pattern on the photoresist is repeated across the entire wafer.

Lithography is similar to film photography, in which a pattern is exposed on photoresist and the photoresist is developed using photographic chemicals. The developing process in both cases washes away the unexposed photoresist (assuming positive resist; for negative photoresist, the exposed photoresist is washed away), leaving the resist in the desired pattern on the wafer's surface. An etching system removes the silicon and other layers that are not covered by the pattern of the photoresist.

In the race to pack more transistors into integrated circuits, manufacturers keep coming up with techniques to reduce the minimum feature size they can print. The method currently used by most high volume integrated circuit manufacturers is called 193 nm immersion lithography. The *193 nm* relates to the wavelength of ultraviolet light generated by a laser used to expose the resist, and *immersion* refers to the fact that you are immersing the lens in a puddle of ultrapure water. Air between the lens and photoresist causes light to bend slightly, due to differences in the index of refraction between air and the lens. However, the index of refraction for water is closer to that of the lens, so the light bends less and the stepper can print a finer pattern. This is the same concept used in microscopes with oil immersion lenses that help you get a higher resolution.

Steppers are built to be used in high-volume manufacturing lines, so they have to process more than 100 wafers an hour and run all day, seven days a week, with minimal downtime for maintenance. A machine capable of this type of resolution, as well as low defect levels and low downtime, costs more than ten million dollars.

When manufacturing integrated circuits, you can expose several different patterns on a wafer and each of these patterns defines a particular layer or type of material. For example, one layer might define the metal lines that connect various components of the circuit, while another layer might define the gate of transistors in the circuit. (The gate of a transistor is the region that allows an applied voltage to turn the transistor on or off and is the smallest region to be patterned in the integrated circuit.) Currently, manufacturers are working with steppers that use 193 nm immersion lithography to produce integrated circuits with a 32 nm minimum feature size.

Although the 193 nm immersion system becomes less inefficient as the feature size is reduced, manufacturers will have to use this system until the next-generation system is available. That next improvement in steppers and lithography will be a system that uses ultraviolet light with a 13.5 nm wavelength. This system is called extreme ultraviolet, or EUV, because it uses ultraviolet light with such an extremely short wavelength. *Extreme ultraviolet nanolithography* systems don't use immersion techniques. Instead, the light path and the wafers that are processed are in a vacuum because air or water would block the EUV beam.

EUV steppers produced thus far have some problems that make them impractical. One issue is that resolutions below 20 nm have not been demonstrated yet because the EUV is intense enough to cause electrons in the photoresist to break free from atoms. These electrons then expose the photoresist, causing the exposed region to grow broader. Also, EUV steppers require a very high-power laser or plasma source to generate enough light to expose a high volume of wafers, and such sources aren't yet available. Current systems can expose only about 15 wafers per hour, whereas production systems need to expose more than 100 wafers per hour to be cost effective.

Writing patterns for researchers: Dip-pen nanolithography

If you're in the market for a low-volume lithography system for your lab, you might try a system that uses *dip-pen nanolithography*. These systems are modified from AFM systems that either obtain images of a surface or deposit ink on that surface. The tip is coated with the ink, and ink is deposited where the tip touches the surface. The ink can be deposited in patterns as fine as 15 nm wide. The width of the pattern depends on how long you leave the tip stationary on the surface; a longer time resting on the surface results in more ink deposited in a wider pattern.

Dip-pen nanolithography systems are useful for making prototypes of nanoscale structures. For example, if you want to see if a molecule will attach to certain other chemical or biological molecules, just coat the substrate with a substance that the first molecule will bond to, for example, gold. Then use the dip pen to deposit the first molecule in an array of dots, as shown in Figure 4-13.

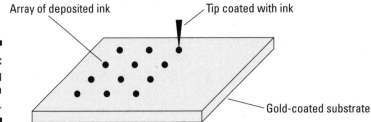

Figure 4-13:
Depositing
ink with a
dip pen.

Array of deposited ink   Tip coated with ink

Gold-coated substrate

After you create the array, you can expose it to the other molecule and then scan the array with an AFM. If an extra molecule is attached, you've succeeded.

E-beam nanolithography

E-beam nanolithography is like ultraviolet-based lithography in that a beam of electrons is focused on the photoresist. However, there are some significant differences between the two methods. For example, electric fields are used instead of a lens to focus the beam. Also, rather than using a mask to define a pattern on the photoresist, the beam moves to create the pattern. The entire operation takes place in a vacuum because air or any other substance blocks the movement of electrons.

You can buy expensive e-beam nanolithography systems (for a few million) or modify the SEM you have in your lab to draw patterns in the resist. Both dedicated systems and retrofitted systems are versatile in producing any pattern a researcher can think of, as demonstrated in Figure 4-14.

Figure 4-14: A pattern created using e-beam nanolithography.

 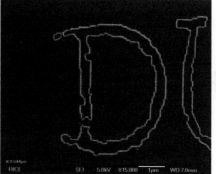

Courtesy of Sungbae Lee at Rice University

Systems designed from scratch specifically for nanolithography can write features as small as 10 nm wide. This capability makes these systems ideal for producing nanoscale patterns in a lab producing prototypes. However, because the e-beam systems have to scan the pattern onto a wafer, rather than stepping a pattern from a prepared mask, they are too slow for high-volume integrated circuit manufacturing.

Nanoparticle growth systems

If you're interested in growing your own nanoparticles, read on! Producing nanoparticles involves breaking up a bulk material into atoms or ions and then allowing those atoms or ions to condense into nanoparticles. This process actually happens, in a basic and an inefficient way, in your campfire, which produces a small quantity of carbon nanotubes.

Equipment designed to produce nanoparticles makes the process more efficient by breaking down more of the raw material to the atomic level under controlled conditions to condense atoms together and form nanoparticles.

Several types of systems are used to produce nanoparticles. A description of a few commonly used systems follows.

In a plasma source system, an inert gas, such as argon, flows into a chamber. This gas carries macroscopic particles of the material from which you want to produce nanoparticles. A high-power radio frequency signal applied to the carrier gas produces plasma, which then flows into a cooled chamber, as illustrated in Figure 4-15. The ions then condense into nanoparticles. This method is often used for volume production of metallic nanoparticles.

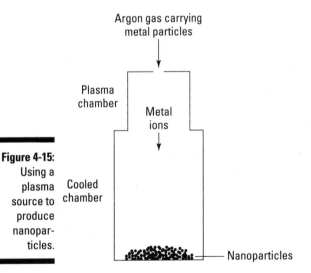

Argon gas carrying
metal particles

Plasma
chamber

Metal
ions

Figure 4-15:
Using a
plasma
source to
produce
nanopar-
ticles.

Cooled
chamber

Nanoparticles

If you're interested in growing only small quantities of nanoparticles, you might try a spark system. This system uses a high-voltage pulse to create a spark between two electrodes. The electrodes are made from the material from which you want to create nanoparticles. The spark causes atoms to evaporate from the electrodes, and the atoms then move into the flow of the argon carrier gas to the condensation chamber, as illustrated in Figure 4-16.

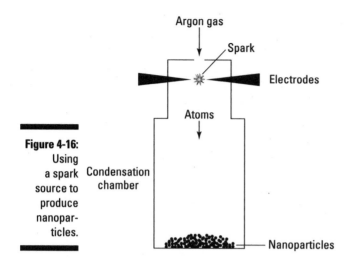

One other option for producing nanoparticles is a laser ablation system. In this type of system, you focus a laser beam on a target that has been pre-heated in a furnace. The target is made of the same material from which you want to create nanoparticles. The laser vaporizes atoms off the target and a flow of argon gas sweeps the atoms downstream to the condensation chamber, as illustrated in Figure 4-17.

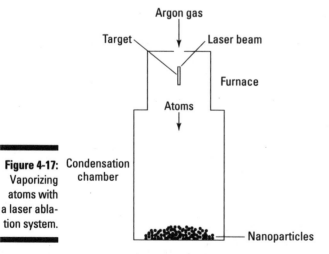

If you want to grow carbon nanotubes, you might also try a plasma-enhanced chemical vapor deposition system. This type of system uses a gaseous source instead of a solid target or powder, as in other systems we've dis-cussed. A gas, such as methane, flows between two electrodes to which you've applied a high-power radio frequency signal. This procedure generates a plasma, breaking up the methane or other source gas into atoms. A substrate is mounted on one of the electrodes, as illustrated in Figure 4-18.

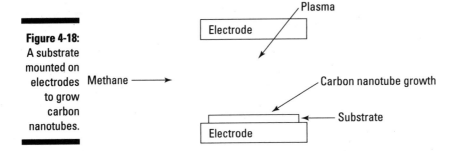

Figure 4-18:
A substrate mounted on electrodes to grow carbon nanotubes.

Particles on the surface of the substrate act like seed crystals because carbon naturally bonds to them. Additional carbon atoms bond to the carbon atoms already in place until — voilá — you've grown a carbon nanotube.

Chapter 5

Putting Nanotechnology to Work

· ·

In This Chapter

▶ Changing the amount of surface area and size of pores in materials

▶ Functionalizing materials to change their properties

▶ Making one from two: nanocomposites

▶ Assembling things with self-assembly and mechanosynthesis

▶ Using nano with mechanical systems and electronics

▶ Integrating nanoparticles into a variety of materials

· ·

*I*n previous chapters, we describe nanotechnology and certain nanomaterials, such as buckyballs and carbon nanotubes. The logical next step is to understand what techniques can be applied to make use of those materials.

The tiny nature of nanoparticles results in some useful characteristics, such as an increased surface area to which other materials can bond in ways that make for stronger or more lightweight materials. Other techniques discussed in this chapter include how nanoparticles can be used in chemical reactions and to create bonds with other materials as well as how to build a variety of things atom by atom.

Changing the Size of Things

At the nanoscale, size does matter when it comes to how molecules react to and bond with each other, so that's our first topic in this section. Secondly, we look at nanopores (very tiny holes), that have their own uses in filtering materials or identifying materials such as DNA strands.

Maximizing surface area

A piece of bulk material (for example, a carrot) contains trillions of atoms, and only a very small portion of its atoms are on the surface. If you take a carrot and cut it in half, you've just increased the surface area. If you cut it into slices, you keep increasing the surface area without increasing the amount of carrot or the number of atoms. However, you have increased the number of atoms exposed on the surface of the pieces.

Think of nanoparticles as being like shredded carrots: They contain anywhere from a handful of atoms up to a few thousand atoms, and have a larger portion of atoms on the surface. The difference in surface area between a bulk material and a sample weight of nanoparticles is illustrated in Figure 5-1.

Figure 5-1:
Increasing
the surface
area with
nanopar-
ticles.

 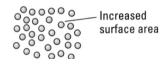

Increased
surface area

Bulk material   Nanoparticles

With a larger number of atoms on the surface, more atoms are available to interact with the atoms or ions of other substances. For example, molecules in an explosive material and oxygen can interact in what's called an exothermic reaction, generating heat. When heat is contained, whatever is holding it in bursts, resulting in an explosion. When the explosive substance is formed of nanoparticles, the increased surface area causes the reaction to occur faster and produces a more powerful explosion because more of the molecules are in contact with oxygen.

This presence of more atoms on the surface of a material can be useful in several applications, including the following:

✔ Improving electrodes used in batteries, resulting in increased charging rate and storage capacity of the battery, as we discuss in Chapter 10.

✔ Improving catalysts to lower the temperature and energy required to run chemical reactions, as we discuss in Chapter 10.

✔ Improving explosives by increasing the power generated by smaller quantities of materials, as discussed in Chapter 12.

Reducing pore size in materials

Porous materials have holes that water or other liquids can seep through. Materials containing nano-sized holes — called nanopores — offer some intriguing possibilities for the way in which pores can be used.

Many substances are porous. For example, porous underground rock (called an aquifer) contains voids that absorb water. However, a significant portion of the volume of the porous rock is solid, limiting the amount of water that it can contain. By making porous material of nanoparticles, the percentage of the material that is solid can be reduced, which allows the nanoporous material to store a gas or liquid within a much higher percentage of its volume.

If a solid substance is riddled with nanopores, you can store a large volume of gas in the solid. A good example is aerogels, which we discuss in more detail in Chapters 7 and 12. Aerogels are made up of nanoparticles separated by nanopores that are filled with air. Because the bulk of the aerogel material (often more than 90 percent of the volume) is air, the material makes a very good insulator.

Another use of reduced pore size using nano is in designing nanoporous materials to maximize the internal surface area and customize the size of the openings. Researchers have created materials called *metal-organic frameworks* (MOFs). MOFs are being developed, for example, to store gases to capture carbon dioxide, as discussed in Chapter 11, or to store hydrogen for fuel cell use, as discussed in Chapter 10.

Having a nanoparticle riddled with nanopores with increased surface area available for contact with gases or liquids can also improve the effectiveness of catalysts. A catalyst disassembles molecules involved in a reaction into atoms. These atoms can then react with the other type of molecule involved in the reaction to form the desired result, an entirely different chemical substance.

Creating MOFs

MOFs are the playground of Omar Yaghi, a chemistry professor at the University of Michigan. Yaghi was attracted to chemistry because of his appreciation for the design of matter. His office reflects that interest: Its décor is largely made up of models of chemical structures. In trying to design chemical structures in the lab, Yaghi became the creator of reticular chemistry. Reticular chemistry is the study of networks of molecules to allow us to use molecular building blocks to assemble structures, previously the work of Nature alone. The professor and his team carefully modify MOFs to obtain the properties they seek. In modifying features, they can make it possible for certain molecules to attach or create larger holes between molecules. One practical outcome of his puttering is increased storage capacity for hydrogen, which could be very useful in fuel cells for devices such as laptops and cars.

When a catalyst is nanoporous, the atoms that are to be disassembled can land on the inside surface of the nanopores, providing more surface area for the catalyst to disassemble the molecules and move the reaction along. These improved catalysts have a better capability to break down air pollutants (as we discuss in Chapter 11), as well as to improve electrodes used in batteries, resulting in increased charging rate and storage capacity of the battery (as we discuss in Chapter 10).

A few more applications of nanopores you can check out elsewhere in this book include the following:

- Membranes containing nanopores are used for water filtration to reduce the energy required to separate water from salts in desalinization of seawater, as discussed in Chapter 11. When the pores are formed with carbon nanotubes, water molecules flow very easily through them, but the larger salt molecules can't pass through.

- You can use nanopores to transport protons (hydrogen ions) in fuel cells, as discussed in Chapter 10. Researchers are developing membranes that include nanopores containing an acidic solution, which makes it easier for the hydrogen ions to pass through the membrane.

- Nanopores can be used to quickly analyze the structure of DNA, as discussed in Chapter 9. When a DNA molecule passes through a nanopore that has a voltage applied across it, researchers can determine the structure of the DNA by the changes in electrical current.

Modifying Material Properties

The properties of nanoparticles can be customized for use in a particular application by bonding molecules to the nanoparticles in a process called *functionalization*. In addition, the capability to build *nanocomposites,* materials formed by integrating nanoparticles into the structure of a bulk material, makes it possible to create new materials that offer a range of new possibilities.

The fundamentals of functionalization

When an atom is attached to another atom, the attachment is called a chemical bond. Functionalization is a process that involves attaching atoms or molecules to the surface of a nanoparticle with a chemical bond to change the properties of that nanoparticle.

The bond used in functionalization can be either a covalent bond or a van der Waals bond. *Covalent bonding,* in which electrons are shared between the atoms, as illustrated in Figure 5-2, involves an atom on the nanoparticle sharing electrons with an atom on the molecule, creating a very strong bond. In a van der Waals bond, electrostatic attraction occurs (negative and positive charges on the molecules and nanoparticles attract each other). A positively charged region of the molecule or nanoparticle and a negatively charged region of the molecule or nanoparticle form a bond. The van der Waals bond is not as strong as a covalent bond, but it also does not weaken the structures being bonded, as covalent bonds do.

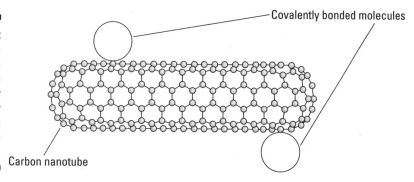

Covalently bonded molecules

Figure 5-2:
Function-
alizing a
carbon
nanotube by
covalently
bonding
molecules
to it.

Carbon nanotube

For example, if you are bonding molecules to carbon nanotubes, a covalent bond might weaken the nanotube while a van der Waals bond would not. Therefore, although covalent bonds are used more often for functionalization, van der Waals bonding is sometimes useful. Figure 5-3 shows one such use: functionalizing a carbon nanotube by bonding a molecule to the nanotube using van der Waals force.

Figure 5-3:
Function-
alizing
a carbon
nanotube by
attaching
a molecule
to it using
van der
Waals
bonding.

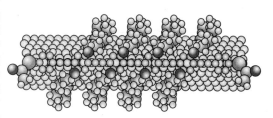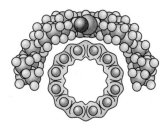

Functionalization is used to prepare nanoparticles for many uses, for example:

- Making sensor elements that can be used to detect very low levels of chemical or biological molecules (see Chapter 6) or for the diagnosis of a blood sample (see Chapter 9).

- Bonding nanoparticles to fibers or polymers to form lightweight, high-strength composites (discussed later in this chapter).

- Making nanoparticles that can bond to biological molecules present on the surface of diseased cells to produce targeted drug delivery agents (see Chapter 9).

- Making nanoparticles that are attracted to prepared attachment sites, such as surfaces containing certain types of atoms (sulfur is attracted to gold, for example) for self-aligned assembly (discussed later in this chapter).

Making nanocomposites

When you include functionalized nanoparticles in a composite material, those nanoparticles can form covalent bonds with the primary material used in the composite. For example, functionalized nanotubes can bond with polymers to produce a stronger plastic. In a carbon fiber composite, functionalized nanotubes bond with the carbon fibers to create a stronger structure, as illustrated in Figure 5-4.

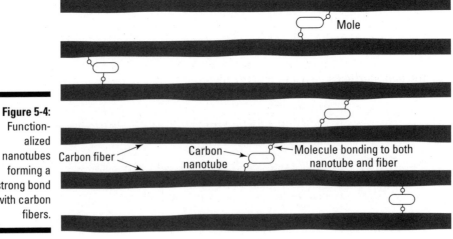

Figure 5-4: Functionalized nanotubes forming a strong bond with carbon fibers.

Mole

Carbon fiber

Carbon nanotube

Molecule bonding to both nanotube and fiber

Nanocomposites are being used in several applications:

✔ A variety of nanoparticles such as buckyballs, nanotubes, and silica nanoparticles are being used with various fibers to form nanocomposites used in sports equipment such as tennis racquets to improve their strength or stiffness while keeping them lightweight (see Chapter 8).

✔ Nanocomposites using carbon nanotubes and polymers are being developed to make lighter-weight spacecraft (see Chapter 12).

✔ Nanocomposites using carbon nanotubes in an epoxy are being used to make windmill blades longer, enabling the windmill to generate more electricity (see Chapter 10).

✔ Nanoparticles of clay are used in plastic composites to reduce the leakage of carbon dioxide from plastic bottles, improving the shelf life of carbonated beverages (see Chapter 8).

✔ Composites of nanoparticles and polymers are being developed to produce lightweight, strong plastics to replace metals in cars (see Chapter 7).

Customizing the Structure of Coatings and Films

Nanoparticles can be included with bulk materials, such as paint, to improve a particular property, for example, making the paint more impervious to ultraviolet rays. The nanoparticle material would be in the form of a powder that is mixed with the paint, just as pigments are added to modify the paint color.

Nanoparticles can also be used to entirely form a film. For example, a nanoparticle that has been functionalized by adding a hydrophobic molecule can be used to create a waterproof film. When you spray a liquid containing the nanoparticles on a surface, the nanoparticles bond to the surface and the hydrophobic molecule points away from the surface to repel water.

The capability to incorporate nanoparticles into coatings and films has several possible applications, including the following:

✔ Walls and other surfaces can be painted with a film containing various types of nanoparticles that kill bacteria. This application can be useful in buildings such as hospitals and doctors' offices (see Chapter 7).

✔ A spray paint for cars is being developed that will turn your car into a solar cell (we discuss this in Chapter 7).

✔ A coating made with aluminum oxide nanoparticles has been shown to dramatically improve the resistance of ship propeller shafts to corrosion (discussed in Chapter 12).

✔ A coating on the inside of tennis balls reduces loss of air and improves their bounce (see Chapter 8).

✔ One coating uses nanoparticles that attach to the surface of porous material, such as stone, giving it waterproof properties (see Chapter 7).

✔ Integrated circuits use films that are only a few nm thick but have sufficient integrity to prevent electrons from leaking through them (see Chapter 6). These films are formed by depositing atoms rather than nanoparticles and must be formed with very uniform thickness and composition even though they are only a few nm thick.

Self-Assembly: Getting Nanoparticles to Make Their Own Arrangements

Some properties of atoms and molecules enable them to arrange themselves into patterns. For example, if you pour a solution containing organic molecules that have a sulfur atom on one end over a gold surface, the sulfur atoms bond to the gold atoms, as illustrated in Figure 5-5. This capability is called self-assembly.

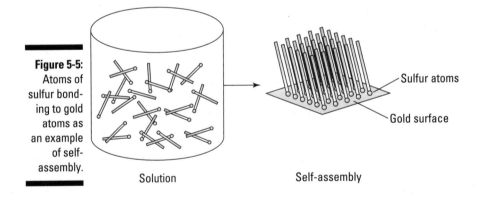

Figure 5-5: Atoms of sulfur bonding to gold atoms as an example of self-assembly.

Solution   Self-assembly

Sulfur atoms

Gold surface

Notice in this figure that the sulfur atoms are all evenly spaced in an array of rows and columns and that the organic molecules standing up from the surface are all leaning slightly to one side. This effect occurs because the sulfur atoms are sharing electrons with the gold atoms in covalent bonds, but the other electrons surrounding the sulfur atoms repel each other. (As you may recall from high school science classes, like charges repel each other.)

This repulsion stops the sulfur atoms from getting too close together. At the same time, the organic molecules are attracted to each other by one of the weaker forces that acts on molecules. This process, called van der Waals bonding, pulls the organic molecules closer together than the sulfur atoms, hence the organic molecules lean slightly to that side.

This mix of covalent bonding, repulsive force, and attractive force results in the molecules, which are functionalized nanoparticles, arranging themselves in a pattern on the gold surface — a perfect example of self-assembly.

Self-assembly is used in many applications, such as

- Building sensors to detect chemical and biological molecules (see Chapter 6)
- Creating computer chips with smaller component sizes, which allows more computing power to be packed on a chip (see Chapter 6)
- Manufacturing diagnostic tools for early detection of diseases (see Chapter 9)

Mechanosynthesis

Researchers are working on building molecular assemblers that use tiny manipulators to precisely position atoms and molecules to build an object from the atoms up (we discuss this topic further in Chapter 4).

When you use millions or trillions of such molecular assemblers in parallel in a process called *massively parallel assembly,* you speed up the building process. The idea is to have the first molecular assembler that you construct build another, and then that new assembler builds another, and so on until there are enough molecular assemblers to build the entire object.

These molecular assemblers are similar to biological machines in your cells called ribosomes. Ribosomes assemble amino acid molecules to construct new proteins, just as molecular assemblers would assemble various atoms or molecules. To differentiate between biologically based molecular assembly and mechanically based molecular assembly, people sometimes refer to the latter as *mechanosynthesis.*

If you want a sneak preview of one person's vision of mechanosynthesis, check out Figure 5-6. This illustration shows a molecular assembly process in which two carbon atoms are deposited on a diamond surface. Various researchers are optimistic about the potential of this nanoassembly process, but it will probably be quite a while before this vision works out.

Positionally-controlled carbon deposition using the DCB6-X dimer placement tool

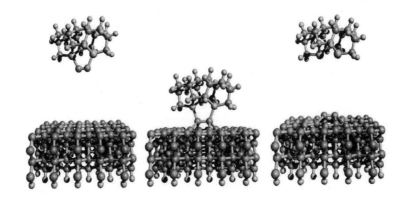

Figure 5-6:
Depositing carbon atoms in a molecular assembly process.

By building an object atom by atom or molecule by molecule, mechanosynthesis can produce new materials with improved performance over existing materials. For example, an airplane strut must be very strong but also lightweight. A molecular fabricator could build the strut atom by atom out of carbon, making a lightweight diamond-like material called a *diamondoid* that is stronger than a diamond.

Remember that a diamond is merely a lattice of carbon atoms held together by bonds between the atoms. By placing carbon atoms, one after the other, in the shape of the strut, such a fabricator could create a diamondoid, which is lighter weight and stronger than any metal.

Current mechanosynthesis research is focused on using carbon atoms to make diamondoid structures. This focus makes sense because, after the method is perfected, we will be able to make useful products such as lightweight airplanes and cars. In the long term, after researchers have perfected mechanosynthesis with carbon atoms, they should be able to expand the technique. They might be able to add selected atoms to carbon that could result in products such as miniscule computer chips and fully functional nanorobots.

In fact, we are carbon-based life forms. Carbon combined with several other atoms make up the biological molecules in our bodies, which just goes to show that carbon combined with other atoms can produce very versatile results.

When mechanosynthesis becomes a reality, it will have applications across society. Although we can't accurately forecast the effect mechanosynthesis will have, here are a few possible examples:

- ✔ Making integrated circuits atom by atom, resulting in much smaller computer chips (see Chapter 6).

- ✔ Building structural components, such as the wings of aircraft, from diamondoids (see Chapter 3). It should be possible to make cars, airplanes, spacecraft, and other objects that weigh a fraction of their current weight.

- ✔ Making molecular replicators that can manufacture most of the material goods you need (see Chapter 8).

- ✔ Making nanorobots that can be used in medicine (see Chapter 9).

The nanorobot shown in Figure 5-7 is designed as a utility nanorobot. For example, if you decide to sit down, you don't have to look for a chair; you just sit down and many of these nanorobots would form together in a lattice with millions of their fellows into the shape of a chair.

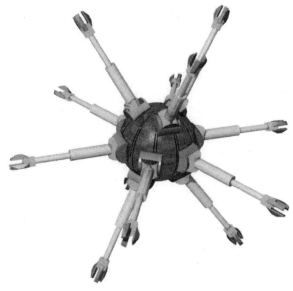

Figure 5-7:
A 12-arm nanorobot which, in swarms, could join to build virtually anything.

Courtesy of J. Storrs Hall

The flexibility of a utility fog

Utility nanorobots, named foglets by the scientist that designed them, J. Storrs Hall, will join in something called a utility fog that could be programmed to make an object appear however you like. For example, you might have a utility fog that forms a hardwood floor, but if you have the urge one day to have a tile floor, they'd shift and become ceramic tile. Utility fogs could also form films, allowing the painted surface of your walls to change color at your whim.

Each utility nanorobot has an antenna arm that you can adjust to reflect and absorb different wavelengths of visible light. If you want an object to be a different color, the nanorobots making up an object adjust their antenna arms to change the color and reflectivity of the object. Although still only a glimmer in scientists' eyes, the concept of utility nanorobots and utility fog could become a reality someday, allowing you to buy one car and yet have a new model every day, or decorate your house in Craftsman style this week and mid-century modern the next. Now that's a useful little utility!

Using NEMs to Work with Nanoscale Gadgets

Nanoelectromechanical systems (NEMS) are devices that integrate nanoscale mechanical and electrical components into a single component. NEMS are built on semiconductor chips using integrated circuit manufacturing techniques.

By manufacturing mechanical components on the same silicon chip that contains electrical circuits used for controlling the mechanism, you can reduce the size, cost, and power requirements for these devices significantly. Currently, the practical version of this methodology is focused on microelectromechanical systems (MEMS), which are used for sensors such as the acceleration sensor in your car that deploys airbags in a crash. (Figure 5-8 shows gears built with this technique.) NEMS offer the potential of increased sensitivity for some types of sensors. We discuss NEMS in more detail in Chapter 6.

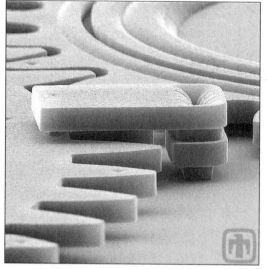

Figure 5-8:
MEMS
gears with
an align-
ment clip
to keep
the gears
meshed.

*Courtesy of Sandia National Laboratories, SUMMiT(TM) Technologies,
www.mems.sandia.gov*

Integrating Nanoparticles into Materials

Weaving materials such as cables with nanoparticles can result in some surprising attributes. In addition, soaking fabrics with nanoparticles can keep
you warmer, drier, or more aromatic, as discussed in this section.

Spinning Nanotubes into Wires and Cables

If you ever read the fairy tale *Rumpelstiltskin,* you remember some poor girl
sitting at a spinning wheel turning wool into gold. Nano researchers actually
want to do something like that with armchair carbon nanotubes (nanotubes
that possess electrical properties similar to metals).

The technique is a little more complicated than that used with wool, but in
this case you end up with a wire with carbon nanotubes aligned in one direction bonded with each other by means of the van der Waals force.

The van der Waals force is an attraction between molecules that forms a
dipole (a positive charge on one end of the molecule and a negative charge
on the other end).

Armchair carbon nanotubes contain some electrons that are free to move within the nanotube. When the nanotubes are placed end to end in a wire, electrons shift on the surface of the nanotubes so that a slightly positively charged end of one nanotube is placed next to a slightly negatively charged end of another nanotube, as illustrated in Figure 5-9. The nanotubes are therefore attracted to each other, which helps to hold the wire together.

Figure 5-9:
Carbon
nanotubes
bonded
together
with van
der Waals
force.

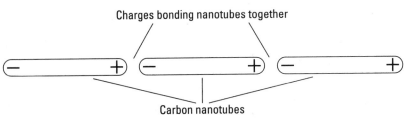

When wires made of armchair carbon nanotubes become available, they could be used to improve materials, such as the following:

✔ Wire with very low electrical resistance: This type of wire could be used to distribute electrical power over much longer distances with fewer losses due to resistance, as discussed in Chapter 10.

✔ Very strong cable: This type of cable could be used in the space elevator to deliver materials into outer space, which is discussed in Chapter 12.

Adding nanoparticles to fibers

When you add nanoparticles to fibers, you can change their properties, making possible things such as lightweight fabric batteries and odor-proof clothing.

A couple of techniques are used to add nanoparticles to fibers. You can simply soak fibers in a liquid that contains nanoparticles, or you can place a positive or negative charge on the fibers and then soak them in a liquid containing nanoparticles with the opposite charge. Because opposite charges attract, the latter method makes it easy to ensure that nanoparticles coat each fiber.

We discuss the applications of nanoparticles in fibers in Chapter 8, including these examples:

- ✔ Fabrics that that repel water and stains by attaching nanoparticles that repel water to the fibers

- ✔ Fabrics with attached nanoparticles that kill bacteria, helping to reduce odors in clothing

- ✔ Nanoparticles that store ions or electrons within fibers, helping to turn fabrics into batteries

Part II
Nano Applications

In this part . . .

Nanotechnology is a science that has applications in almost every area of life, from healthcare to manufacturing, space travel to improving our environment. Keeping up with the constantly evolving advances is a challenge in a field that spreads into so many application areas. But without understanding how nano is being used and applied, understanding the possible depth of its effect on your life is difficult.

In this part, we explore what's being done, developed, or just imagined in the area of nanotechnology today in various industries and settings. With applications such as reversing aging and advancing space travel, we're betting the exciting possibilities we discuss in the chapters in this part may just leave you a bit breathless.

Chapter 6

Nanoscale Electronics

*N*anotechnology will play a key role in transforming laptops as well as other electronic devices over the next few years. Who knows what the laptop of tomorrow will be? It could weigh just ounces and run for weeks on a single charge. Imagine impressing your friends with a laptop that includes a display that unfurls like a sail and a case that comes in every color of the rainbow. At some point, computer chips will become powerful enough that a device the size of your cell phone — or even your wristwatch — will have as much computing power as the most powerful desktop computer available today.

Nanoelectronics encompasses electronic devices that contain components with features of less than 100 nanometers in size. From making better computer chips and improving computer memory to improving displays and providing nanosensors, nanoelectronics has many advantages to offer. That's what we explore in this chapter.

Working with Computer Chips

As one of the authors of this book who worked at Intel for 13 years can tell you, the semiconductor industry has been in the nanotechnology business for years. They use tools and processes to etch nano-sized patterns on silicon wafers. Those patterns make up the circuits on the chip that allow your computer to process data.

The process used to make these patterns is called nanolithography, which we discuss also in Chapter 4.

Microprocessor manufacturers are using nanolithography to make processors with nanoscale transistors that use less power and fit more transistors on each silicon chip, thereby providing higher computer performance.

In this section, we discuss how improvements in nanolithography and changes in the nanoscale structure of the transistors are increasing the density of transistors in microprocessors.

Microprocessors with a 32-nanometer minimum feature size have as many as 995 million transistors packed on one computer chip. Today's microprocessors are a major change from microprocessors in the early 1970s, which had a minimum feature size of 10 micrometers (a micrometer is a thousand times bigger than a nanometer) and 3500 transistors. The current generation of microprocessors is being built with 32-nanometer-gate-width transistors, and processors that use 22-nanometer-gate-width transistors should be available soon, as the race to increase the computing capabilities of your laptop continues.

Seeking smaller chips

Lithography is similar to film photography in that a pattern is exposed on photoresist and the photoresist is developed using photographic chemicals. The developing process in both cases washes away the photoresist that was exposed (assuming positive resist; for negative photoresist, the unexposed photoresist is washed away), leaving the resist in the desired pattern on the surface of the wafer.

By continuing to reduce the minimum feature size that can be created through nanolithography, computer chip manufacturers can do one of the following:

- Increase the number of transistors that can be packed into a microprocessor, which increases the computing power of the device
- Shrink the size of a computer chip with the same number of transistors, reducing the cost of manufacturing the chip

However, there are some limitations to how small a minimum feature size this top-down method can produce. As researchers keep finding innovative ways to make the minimum feature size smaller and smaller, and to increase the number of transistors per chip, they will someday hit a wall. At some point below a minimum feature size of 10 nm, instead of using the top-down method of etching away silicon, it'll probably be necessary to use nanoparticles of the desired size to form these minimum features; or it may even be necessary to use bottom-up methods to build part or all of the integrated circuit from individual atoms or molecules.

We discuss the difference between top-down and bottom-up approaches to nanotechnology in Chapter 1.

Switching things with FETs

Shrinking the transistor is key to making your computer more powerful. The structure of the type of transistor used in microprocessors containing hundreds of millions of transistors on an integrated circuit is called a FET, for field-effect transistor. A FET is shown in Figure 6-1.

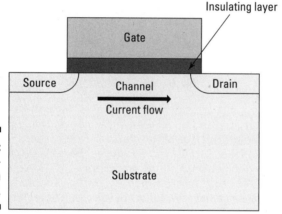

Figure 6-1: The structure of a FET.

Placing a voltage on the gate allows a current to flow through the channel between the source and the drain. The transistor is therefore acting as a switch, not unlike the wall switch you use to turn your lights off and on. Current flows when voltage is applied to the gate; current stops flowing when no voltage is applied to the gate. As the channel length gets smaller, however, the chance of current leaking through the channel between the source and the drain increases, even when no voltage is on the gate.

Integrated circuit manufacturers are planning to modify this structure for minimum feature sizes of about 14 nm and less to reduce the amount of leakage through the channel. This modified transistor, called a finFET because of the fin-shaped channel above the substrate, is shown in Figure 6-2.

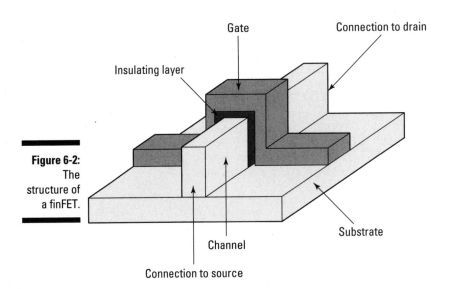

Gate

Connection to drain

Insulating layer

Figure 6-2:
The
structure of
a finFET.

Substrate

Channel

Connection to source

With the gate on the top and two sides of the channel, the voltage applied to the gate has more effect on the channel than in the conventional FET, which has the gate only on the top surface of the channel. Researchers report that, compared to conventional FETs, this setup reduces the leakage of current through the channel when you turn off the transistor.

As we go to press, Intel has announced that they are implementing a finFET transistor structure called Tri-Gate on their 22-nm microprocessors. Using these transistors will provide either reduced power consumption at the same speeds as their current 32-nm microprocessors or increased speeds with the same power consumption.

Using a nanowire as the channel of the FET is a method that researchers are exploring to make even more progress in reducing current leakage. A nanowire transistor consists of a nanowire made of semiconducting material, such as silicon, connecting the source and drain of the transistor, with a gate controlling the current flow through the nanowire. Figure 6-3 shows the structure of a nanowire transistor; the nanowire is vertical, like the fin, rising up from the substrate.

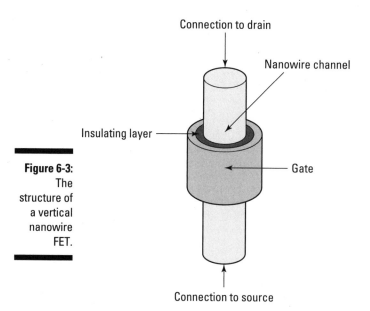

Connection to drain

Nanowire channel

Insulating layer

Gate

Figure 6-3:
The
structure of
a vertical
nanowire
FET.

Connection to source

Using a nanowire as the channel allows you to completely wrap the gate around the channel. This should allow the voltage applied to the gate to have even more control over the channel than when using the finFET (refer to Figure 6-2). This vertical structure also saves space, allowing a higher density of transistors on a chip. Millions or billions of vertical nanowires could be grown on a substrate, like a dense but tiny forest.

Working at the atomic level with mechanosynthesis

Using mechanosynthesis, which is discussed in more detail in Chapter 5, to build integrated circuits is an interesting long-term prospect. The idea is to build integrated circuits atom by atom. In this process you use carbon atoms to make a diamond-like structure with other types of atoms placed as needed to give that part of the structure the semiconductor or conducting properties required to create transistors and other circuit elements. Using this process could result in much smaller computers by reducing the size of a computer chip by about a million times. If this reduction in chip size comes to pass, it will be easy to fit in a wristwatch-sized device all the processing power you need. This smaller chip of the future won't happen in the next few years, but it will be interesting to see where this idea has led in a few decades.

Packing transistors in

Researchers are investigating a variety of other nanomaterials to make smaller transistors and to pack them more tightly in integrated circuits. Two of the leading contenders are:

- ✓ **Quantum dots:** Researchers have demonstrated that they can build transistors in which quantum dots form the channel through which current flows. That channel can be as small as 4 nanometers in diameter. The challenge here is to develop a method of integrating these nanoparticles into a process to build very dense integrated circuits.

- ✓ **Carbon nanotubes:** You can use semiconductor-type carbon nanotubes as transistor channels, similar to the way you can use nanowires. The use of carbon nanotubes, however, has various complications. For example, when you grow carbon nanotubes, both semiconductor- and metallic-type carbon nanotubes are produced. (We discuss these types of nanotubes in Chapter 3.) You need a step in the manufacturing process that runs a current through the metallic nanotubes to intentionally burn them out, just as you might unintentionally burn out a fuse. Researchers must work out details like this before they can use carbon nanotubes in mass-produced integrated circuits.

Improving Your Memory

Hard drives provide the standard method of storing information on a computer. Hard drives require a motor to spin their magnetic disk. Because of this spinning motion, hard drives consume more power and have more chance of failure than memory components that don't have any moving parts.

For that reason, solid-state drives have become popular on smaller computers, such as the iPad. These solid-state drives take up less space, use less battery power, access data faster, and are more hardy (and therefore less likely to be damaged if the device is dropped).

Getting flash-y

Solid-state drives currently in use store information on a type of transistor called flash. The structure of a flash transistor is shown in Figure 6-4. Currently, flash memory manufacturers use nanolithography techniques to build memory chips with minimum feature sizes as small as 20 nm.

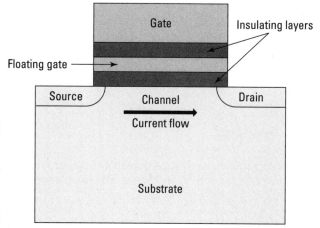

Figure 6-4:
The
structure of
a flash
transistor.

The flash transistor structure is like FETs used in microprocessors (described in the previous section), with a floating gate added. Here's how the flash transistor works as a memory device:

✔ When electrons are stored on the floating gate, they stay there, even if the power is turned off, because the floating gate is surrounded by an insulating material that traps electrons.

✔ Electrons on the floating gate repel electrons from the surface of the channel between the source and the drain, preventing current from flowing through the transistor.

✔ When the computer reads data from the flash memory, it reads ones (current flows through the flash transistor because electrons are not stored on the floating gate) or zeros (current doesn't flow through the flash transistor because electrons are stored on the floating gate).

Due to competition between flash memory manufacturers, they will continue to reduce the minimum feature size so they can increase the amount of information stored on these solid-state drives, whether they are used as the main memory for tablets or laptops or as accessory memory devices such as USB memory sticks. Researchers are using nanotechnology to create other types of computer memory, attempting to leapfrog flash memory in the marketplace. Various companies and universities are developing four methods of using nanowires or nanoparticles to increase the amount of memory stored on solid-state drives, described in the following sections.

Making memory with memristors

Hewett Packard is developing a memory device that uses nanowires coated with titanium dioxide. One group of these nanowires is deposited parallel to another group. When a perpendicular nanowire is laid over a group of parallel wires, a device called a *memristor* is formed at each intersection.

A memristor can be used as a single-component memory cell in an integrated circuit. By reducing the diameter of the nanowires, researchers believe memristor memory chips can achieve higher memory density than flash memory chips.

If I had a nickel . . .

Magnetic nanowires made of an alloy of iron and nickel are being used to create dense memory devices. Researchers at IBM have developed a method to magnetize sections of these nanowires. By applying a current they can move the magnetized sections along the length of the wire. As the magnetized sections move along the wire, the data is read by a stationary sensor. This method is called race track memory because the data races by the stationary sensor. The plan is to grow hundreds of millions of U-shaped race track nanowires on a silicon substrate to create low-cost, high-density, and highly reliable memory chips.

Using silicon dioxide sandwiches

Another method of using nanowires is being investigated at Rice University. Researchers at Rice have found that they can use silicon dioxide nanowires to create memory devices. The nanowire is sandwiched between two electrodes. By applying a voltage, you change the resistance of the nanowire at that location. Each location where the nanowire sits between two electrodes becomes a memory cell.

The key to this approach is that researchers have found that they can repeatedly change the state of each memory cell between conductive and nonconductive without damaging the material's characteristics. These researchers believe that they can achieve high memory densities by using nanowires with a diameter of about 5 nm and by stacking multiple layers of arrays of these nanowires like a triple-decker club sandwich.

Getting magnetic

An alternative method being developed to increase the density of memory devices is to store information on magnetic nanoparticles. Researchers at North Carolina State University are growing arrays of magnetic nanoparticles, called nanodots, which are about 6 nm in diameter. Each dot would contain information determined by whether or not they are magnetized. Using billions of these 6-nm diameter dots in a memory device could increase memory density.

It will be interesting to see how these methods, and the work by existing memory manufacturers to improve existing memory storage devices, pan out. Which type of memory devices we'll be using 5 or 10 years from now will come down to the right mix of memory density, power consumption, data read and write speeds, and which technique attracts the funding to build manufacturing plants.

To keep up with the latest advances in this area, visit the companion web site at `www.understandingnano.com/nanotechnology-computer-memory.html`.

Connecting with Light

You've seen how the speed of your connection can have a big effect on your Internet browsing or movie downloading experience. One factor that affects your connection is the type of cable that the data is carried on. The cable that enables the fastest connection is fiber-optic cable, which carries data as light, as opposed to copper wire, which carries data as electrons.

Because of the superiority of this connection method, fiber-optic cable has been installed between cities, connecting major buildings in each city to provide the fastest possible transmission over the main data trunk.

Researchers are planning to use nanoscale components to adapt this methodology for carrying data within computers. The idea is to use light to carry data between microprocessor cores within a computer chip and between separate chips within a computer.

Today, microprocessors have a certain number of cores, from one to eight or more. Multiple core processors allow several mathematical or logic calculations to run at the same time. Within the microprocessor cores are connections between components, such as transistors. Nano could replace the current technology, which sends data through metal lines, with metallic carbon nanotubes, which conduct electricity better than metal. When you need to send information from one core to another, the outgoing electrical signal would be converted to light and travel through a waveguide to another core, where a detector would change the data back to electrical signals.

In addition to speeding up data transmission between cores in a microprocessor, this method might also lower power consumption. That savings occurs because all metal wires have a resistance to the movement of electrons through them, so some of the voltage used to drive the electrons is converted to heat. Researchers have developed techniques for transmitting light that customize the nanostructure of crystalline material to form waveguides. These waveguides allow light of a particular wavelength to travel through the material with almost no loss of energy.

Researchers are developing nanoscale light sources, electrically driven optical switches (also called modulators), waveguides, optical routers, and detectors to convert electrical data into optical data, route it to a microprocessor core, and convert the optical data back to electrical data so that the microprocessor core can then process it.

Because physicists often use the term *photons* when referring to light, using nanoscale components or nanostructured materials to manipulate light is called *nanophotonics*.

One nanoparticle-based method of generating light, developed at Cornell University, is called a *spaser* (surface plasmon amplification by stimulated emission of radiation). A spaser is similar to a laser. The difference between a spaser and a laser is that a laser has a cavity in which light bounces back and forth to amplify the light intensity in a process similar to resonance. That method won't work very well with a nano-sized light source, whose size is a fraction of the wavelength of the light you're trying to generate. A spaser is much smaller than the wavelength of light; in fact, the spaser made at Cornell is made of a 44-nm diameter particle, as shown in Figure 6-5, and generates light with a wavelength of 531 nm.

Figure 6-5:
A nanoparticle that works like a laser.

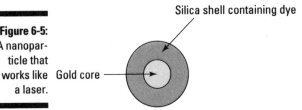

Silica shell containing dye

Gold core

When you excite dye molecules (you can use either light or electrical signals to do this) in the outer shell of the spaser, the dye molecules add electrons to the gold core. These electrons, along with electrons in the conduction band of the gold core, form an electron cloud (called a *plasmon*). This cloud oscillates on the surface of the gold core at the same frequency as the wavelength of light that you want to generate. These oscillating electrons generate an electric field that is strengthened by the resonance oscillation and additional electrons supplied by the dye molecules until the spaser generates a pulse of light.

In another approach, IBM is developing a carbon nanotube–based laser as a light source. Carbon nanotubes generate light; the wavelength of that light depends on the diameter of the nanotube. Either an electrical signal or a light signal can be used to get a nanotube to initiate light generation. The nanotubes are located between two mirrored surfaces; the distance between them is half the wavelength of the light being used. These mirrored surfaces act as the resonance cavity of the laser, which amplifies the light generated by the nanotubes.

Improving Displays

Could a laptop computer display unroll like a portable movie screen, or could you detach it from your laptop and attach it to the back of an airline seat with Velcro? These are just some of the ways that nanotechnology might change the physical form of laptops.

Using nanowires in OLEDs

Researchers at the University of Michigan have demonstrated that they can use nanowires as electrodes in organic light emitting diode (OLED) displays, which makes it possible to manufacture larger and more flexible OLED displays.

One option to nanowire-enabled flexible OLED displays could be a very thin, low-power, high-resolution screen that uses nanotubes. Various companies have worked on such a display, called a nano-emissive display because the nanotubes emit electrons at each spot on the display that must be illuminated to form a picture. This display actually works much like an old-fashioned television set but can provide laptops with very lightweight screens and fine enough resolution for a high-definition TV display.

Making displays flexible

The display in your cell phone and computer includes layers of transparent, conductive material called indium tin oxide. These layers of indium tin oxide have two problems: this material is expensive, and it cracks if you flex it. To protect these layers, they're mounted on a rigid glass substrate. Unfortunately, growing a film of indium tin oxide on a substrate involves an expensive vacuum deposition process.

The folks designing new electronic gadgets like the idea of flexible display screens. They think that, once the display screens become flexible enough, they can make gadgets such as flexible cell phones that can be rolled into the shape of a bracelet that you can wear around your wrist.

To satisfy the desire for flexible displays and reduced costs, researchers are investigating nanomaterials that are conductive, transparent, flexible, and less expensive to apply. Silver nanowires deposited on plastic sheets are a promising option because the silver provides electrical conductivity, has about the same optical properties of indium tin oxide anodes, and is flexible.

Another option that can achieve flexibility, electrical conductivity, and transparency is using sheets of metallic carbon nanotubes. These nanotube-based sheets are transparent because carbon nanotubes absorb infrared light rather than visible light. As we discussed in Chapter 3, carbon nanotubes can be bent, therefore these thin (about 50 nm thick according to one vendor) sheets of carbon nanotubes are ultra-flexible.

Carbon nanotubes are still fairly expensive because large-scale nanotube manufacturing facilities have only recently been built, so the material isn't widely available. Also, to manufacture a highly conductive sheet, you need to separate metallic-type nanotubes from semiconductor-type nanotubes, a complication that adds to the cost.

A third option to get the desired flexibility is to use graphene. Like carbon nanotubes, graphene does not absorb visible light, does conduct electricity, and is flexible. Graphene is expensive, however, because methods to separate graphene sheets, which are only one atom thick, from graphite (the bulk form of graphene), were only discovered a few years ago. As a result, graphene is not yet manufactured in quantities that make it cost effective.

For the near future, silver nanowires will be the preferred method used to produce flexible display screens. As the cost of carbon nanotubes or graphene comes down, one of those materials may take the lead in the marketplace.

Reducing power consumption in quality display screens

You may love those high-definition big-screen TVs, but they consume lots of power; even the display screen on your laptop consumes a sizable portion of your laptop's battery power. Although integrated circuit manufacturers have worked to reduce the power consumed by the microprocessor in laptops, the next big opportunity to reduce power usage is the display screen. Nanotechnology may be able to reduce the power consumed in both high-definition big-screen TVs and computer displays in a couple ways.

Using carbon nanotube emissions

Old-style bulky TVs and computer screens used a cathode ray tube that scanned a beam of electrons across the TV screen and caused fluorescent dots on the screen to light up, displaying an image. Display screens using carbon nanotubes also light up fluorescent dots with not one source of electrons but a set of carbon nanotubes for each pixel. (Each pixel has a green, a red, and a blue fluorescent dot. For each of those dots, a nanotube emits electrons that combine to produce the specific color for that pixel.)

When a pixel must be lit up to form a picture, the nanotubes for that pixel emit electrons. This method, called *nanotube emissive display,* provides high-definition images using less power. Making nanotube emissive displays, however, is challenging. For example, these displays require a high-quality vacuum in the gap between the tip of the nanotubes and the fluorescent dots that will last for the lifetime of the display. The vacuum lets electrons travel between the tips of the carbon nanotubes and the fluorescent dots. In addition to the need for a high-quality vacuum, these nanotubes are expensive. At this point, various companies have attempted to develop displays using nanotube emissive technology but have dropped the projects or sold their technology to other companies.

Taking advantage of quantum dots

To excite the fluorescent dots used in many displays, you need to shoot electrons at pixels. Quantum dot displays, on the other hand, use an electric field to excite electrons and generate light. You control the light generated in each pixel by using electrodes to apply a voltage.

Quantum dot displays should be simpler to make as well as use less power. However, manufacturers must overcome one issue. To make color displays using three different colors, you need three different diameter quantum dots. You essentially have to put three different sizes of quantum dots in each pixel, which is a difficult technique to master. Research published recently shows a method created by one company to get around this issue, but only time will tell if this technique will become practical.

Detecting All Kinds of Things with Nanosensors

Nanotechnology can enable sensors to detect very small amounts of chemical vapors. Various types of detecting elements, such as carbon nanotubes, zinc oxide nanowires, or palladium nanoparticles, can be used in nanotechnology-based sensors. These detecting elements change their electrical characteristics, such as resistance or capacitance, when they absorb a gas molecule.

Because of the small size of nanotubes, nanowires, and nanoparticles, a few gas molecules are sufficient to change the electrical properties of the sensing elements. This characteristic allows these particles to detect a very low concentration of chemical or bacterial molecules in the air. The goal is to create inexpensive *nanosensors* that can sniff out hazardous substances just as dogs are used in airports to smell the vapors given off by explosives or drugs.

Scoping out the potential of nanosensors

The capability to produce small, inexpensive sensors that can quickly identify a chemical vapor gives us a kind of nano-bloodhound that doesn't need sleep or exercise and that could be useful in a number of ways beyond airport security:

- ✔ **Industrial settings:** These sensors could be useful in industrial plants that use chemicals in manufacturing settings to detect the release of chemical vapors.

- ✔ **Sensing hydrogen leaks:** When hydrogen fuel cells come into use, in cars or other applications, a sensor that detects escaped hydrogen could warn of a leak.

- ✔ **Air quality monitoring:** This technology should make it possible to build inexpensive networks of air quality monitoring stations to improve the tracking of air pollution levels.

- ✔ **Mobile phones:** With the technology applied to mobile devices, a food science worker out in the field could wave a cell phone over a head of lettuce and have the device sense the presence of salmonella or dangerous levels of pesticides, for example.

- ✔ **Contagious airborne diseases:** You could use nanosensors to scan people boarding planes to stop disease carriers from travelling and reduce the spread of pandemic contagions.

How nanosensors do what they do

Manufacturers are currently using nanomaterials to build sensors in several ways, including the following:

- ✔ **Sensors using semiconductor nanowire detection elements:** These sensors are capable of detecting a range of chemical vapors. When molecules bond to nanowires made from semiconducting materials such as zinc oxide, the conductance of the wire changes. The amount that the conductance changes and in which direction depends on the molecule bonded to the nanowire. For example, nitrogen dioxide gas reduces how much current the wire conducts, and carbon monoxide increases the conductivity. Researchers can calibrate a sensor to determine which chemical is present in the air by measuring how the current changes when a voltage is applied across the nanowire.

✔ **Semiconducting carbon nanotubes:** To detect chemical vapors, you can first functionalize carbon nanotubes by bonding them with molecules of a metal, such as gold. Molecules of chemicals then bond to the metal, changing the conductance of the carbon nanotube. As with semiconducting nanowires, the amount that the conductance and its direction changes depends on the molecule that bonds to the nanotube. This type of sensor is now commercially available.

✔ **Carbon nanotubes and nanowires that detect bacteria or viruses:** These materials can be used also to sense bacteria or viruses. First you functionalize the carbon nanotubes by attaching an antibody to them. When the matching bacteria or virus bonds to an antibody, the conductance of the nanotube changes. In this process you attach nanotubes to metal contacts in the detector and apply a voltage across the nanotube. When a bacteria or virus bonds to the nanotube, the current changes and generates a detection signal. Researchers believe that this method should provide a fast way to detect bacteria and viruses. One promising application of this technique is checking for bacteria in hospitals. If hospital personnel can spot contaminating bacteria, they may be able to reduce the number of patients who develop complications such as staph infections.

✔ **Nanocantilevers:** These devices are being used to develop sensors that can detect single molecules. These sensors, made with NEMS techniques, take advantage of the fact that the nanocantilever oscillates at a resonance frequency that changes if a molecule lands on the cantilever, changing its weight. Coating a cantilever with molecules, such as antibodies, that bond to a particular bacteria or virus determines what bacteria or virus will bond to the cantilever. Figure 6-6 illustrates nanocantilevers; here virus particles bond to antibodies on the cantilevers.

 Coincidently, our intrepid technical reviewer for this book works at the Birck Nanotechnology Center, two of whose researchers provided us with Figure 6-6. Birck is at the leading edge of nanoelectronics research, and you can read more about them by visiting their web site at www.purdue.edu/discoverypark/Nanotechnology/.

One example of nanoparticles used in sensors is a hydrogen sensor that contains a layer of closely spaced palladium nanoparticles that are formed by a beading action similar to water collecting on a windshield. When hydrogen is absorbed, the palladium nanoparticles swell, which causes shorts between nanoparticles and lowers the resistance of the palladium. This process is illustrated in Figure 6-7. Another use of nanoparticles is in the detection of volatile organic compounds (VOCs). Researchers have found that by embedding metal nanoparticles made of substances such as gold in a polymer film, you create a VOC nanosensor.

Figure 6-6:
Viruses
bonding to
antibodies
on nanocan-
tilevers.

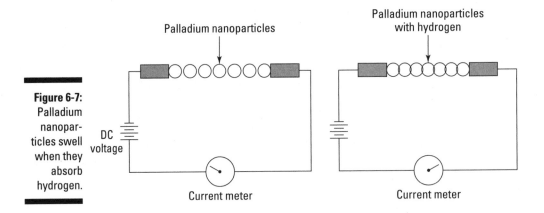

Figure 6-7:
Palladium
nanopar-
ticles swell
when they
absorb
hydrogen.

Sensors are also being developed to detect molecules that indicate that a particular disease is present in a blood sample. These diagnostic sensors are discussed in Chapter 9.

NEMS

NEMS, or *nanoelectromechanical systems,* are devices that integrate nanoscale mechanical and electrical components into a single component. NEMS are built on semiconductor chips using integrated circuit manufacturing techniques. By manufacturing mechanical components on the same silicon chip that contains electrical circuits used for controlling the mechanism or amplifying a signal, you significantly reduce the size, cost, and power requirements for these devices.

Currently, the practical version of this methodology is focused on *microelectromechanical systems* (MEMS), which are used for sensors. An example of a MEMS sensor is the acceleration sensor in your car that deploys airbags in a crash. MEMS are also used in game controllers to sense your movements and then use that data to control an action figure in a game. Medical applications for MEMS include blood pressure sensors or pumps to sense low blood sugar and deliver insulin.

NEMS can make some types of sensors more sensitive. Researchers believe that moving to NEMS, which reduces the size of the mechanical components on a chip to the nanoscale, can improve the capabilities of devices that need extremely sensitive measurements.

Researchers have developed chemical and bacterial NEMS sensors using nanocantilevers (as discussed in the preceding section). Nanocantilevers have been fabricated with thicknesses as low as 20 nm. These act like a guitar string that vibrates at a certain resonance frequency when plucked. Nanocantilevers have been functionalized with antibodies that bond to viruses or bacteria, and the added weight of the virus or bacteria changes the frequency at which the cantilever vibrates.

Researchers have developed a NEMS resonator that is sensitive enough to measure the mass of individual molecules. The resonator consists of a ribbon 100 nanometers thick and 2 micrometers long. When a molecule lands on the ribbon, the frequency of its oscillation changes; the amount of the change depends on the mass of the molecule. Because methods to manufacture NEMS devices use the same techniques for manufacturing integrated circuits, researchers envision NEMS devices that have an array of these resonators, just as integrated circuits have arrays of memory cells. Arrays of resonators that measure the mass of many molecules at the same time could be useful in understanding how biological systems work so we can build systems similar to them.

Chapter 7

Nanotechnology in Your House and Car

*N*anotechnology may not (yet) be used to build better mousetraps, but it is definitely involved in building better houses and cars, as well as the electronic gadgets that populate homes and the solvents used to clean them.

In this chapter, we cover ways that nanotechnology is helping to build stronger and warmer houses; to make your computer and other electronic devices smaller and faster; to create cleaning products that kill bacteria and clean better; to protect your car from chips, dents, and rain; and to provide more efficient car batteries.

Building Tougher Building Materials

The construction industry has much to gain from nanotechnology. Solutions in the offing range from materials with better insulating properties to solar cells that power your house more economically to siding that is protected from the effects of weather.

Insulating windows

Windows may provide an interesting view, but they are one of the weakest points in buildings as far as heat loss is concerned. Various manufacturers are trying to reduce that loss with creative new nanoproducts made of nanoparticles that are separated by nanopores filled with air. Stagnate air is a very poor conductor of heat, so conventional insulating material, such as fiberglass, traps pockets of air in the walls of your house to prevent heat from entering or leaving the house. In nano products, a much larger portion of the insulating material is composed of trapped air pockets, making for a better insulator.

The rating that tells you how much a window insulates from heat conduction is the R-value; the higher the R-value, the more a window stops heat from escaping. Double-pane windows are rated at about R3, while exterior walls in houses generally contain insulation that is rated at R19. (These numbers vary somewhat depending on the region of the country.)

One company that's improving the situation is Cabot Corporation, which has developed an aerogel product that reduces heat conduction through windows. Cabot has teamed up with a window manufacturer, Advanced Glazings Ltd, to include Nanogel, a translucent silica aerogel, as an insulating layer in double-pane windows, giving the windows a rating of R17. The web site for Advanced Glazings (www.advancedglazings.com) is shown in Figure 7-1. These windows are translucent, rather than clear, but translucent windows are acceptable or even preferred in plenty of places, and these windows could have a significant effect on heating or cooling costs.

Figure 7-1: Advanced Glazings web site.

Creating thinner walls that hold in heat

Aerogels using nano have other applications in construction, where they can be used to insulate buildings. For example, Aspen Aerogel says that 1 inch of their Spaceloft insulation will give an R-value of R10.3. Conventional insulation must be 5.5 inches thick to get an R19 insulating value, but you get more insulating bang than R19 with only a 2-inch layer of Spaceloft.

One use of this product is to add insulation to existing houses without making the walls significantly thicker, which would reduce the living space. For example, the Proctor Group (www.proctorgroup.com/) is using Spaceloft in their Spacetherm insulation. Figure 7-2 shows a project in which they added only 40 mm (about 1.6 inches) of Spaceloft to the inside of uninsulated brick walls to create much more efficient insulation. Cabot Corporation can supply their aerogel material, Nanogel, in granules, which can allow builders to use the insulating material to fill wall cavities, as well as to install it in sheets.

Figure 7-2: Spacetherm insulation offering improved insulation with less bulk.

Another use for aerogel is to block thermal loss through the metal or wooden studs that provide the framework for a house. Yet another aerogel manufacturer, Thermablok (www.thermablok.com), illustrates the way that their product is installed in a typical house on their web site, as shown in Figure 7-3.

Peel & Stick Adhesive Backing

Install on All Framing Edges

Thermablok Thermal Strip Insulation

Thermablok may be installed either on the exterior or interior side of the wall framing. Insulate the full length of framing members. Apply to all wall, floor or ceiling framing edges: headers, footers, trusses, window and door frames, floor joist and roof rafters.

It is suggested that Thermablok be applied to the exterior side of the assembly, if:

- You have access to entire building envelope, between floor joists and other areas not accessible from the inside.
- With metal framing you are able to create a warm stud, preventing condensation.

Cut and fit Thermablok 1-1/2" wide encapsulated strips tight to any protrusions or interruptions to the insulation plane. Butt edges and ends of Thermablok thermal bridging barrier insulation tightly to each other without gaps.

Remove release backing from adhesive tape on Thermablok thermal bridging barrier insulation, and apply directly to either metal or wood framing. For wood framing you may use a staple hammer with applicable length staples.

Immediately following application of self-adhering Thermablok, press thermal bridging barrier strip firmly and uniformly to substrate surfaces to insure adhesion.

Install covering materials wall, roof, sheathing, plaster lathe or gypsum wallboard in accordance with guidelines.

Plan and Coordinate the Thermablok aerogel insulation work to minimize the generation of off-cuts and waste. Reuse Thermablok off-cuts to the maximum extent possible.

Do not install self-adhering Thermablok thermal bridging barrier insulation when temperature or weather conditions are detrimental to successful installation.

Figure 7-3:
Houses can lose heat through wooden studs, but Spacetherm cuts the loss.

Powering your house with inexpensive solar cells

Various companies are using nanotechnology to develop low-cost solar cells by replacing the expensive semiconductor crystalline starting materials with less expensive materials such as semiconductor nanoparticles or organic molecules. Nanotechnology also helps lower the expensive high-temperature manufacturing processes used to make conventional solar cells by using less expensive manufacturing methods such as ink jet printing techniques. (This process is discussed in detail in Chapter 10.) These companies are working to put solar cells not just on the roof of your house but also in your windows, skylights, and siding.

Incorporating solar cells into buildings is referred to as building integrated photovoltaics (BIPV).

One area where solar cell companies are focusing is integrating solar cells into windows. Global Photonic Energy Company (www.globalphotonic. com/) created ClearPower technology, which they use to produce almost completely transparent windows that can generate solar power. A thin nano-film on a window contains several transparent electrical contacts. The film in ClearPower windows absorbs much more light by size than more traditional solar cells, and you can adjust the color of the film.

Konarka Technologies (www.konarka.com) might allow buildings to generate their own power when their Power Plastic is packaged in glass products such as windows and skylights. The product can both capture solar power and protect people inside the house from harmful effects of the sun.

SolarmerEnergy (www.solarmer.com/productbipv.php) offers plastic solar cells that are easy to install and work well in weaker light. These cells can generate electricity at about 20 percent of the cost of ordinary solar cells and can be built right into building materials.

You will see solar cells incorporated into other portions of buildings than just windows and roofs. At least one company is interested in incorporating its solar cells into garage doors, skylights, walls, decorations, facades, tiles, and shingles.

Protecting siding with nano coatings

Companies are using nanoparticles of UV-absorbing substances such as zinc oxide in clear, protective coatings for wood that allow the wood to show through. Several companies, such as Antaria (www.antaria.com) and Nanophase (www.nanophase.com), are producing such products. For example, Antaria, whose web site is shown in Figure 7-4, is using nanoparticles to make a clear wood coating that protects the wood from discoloration due to ultraviolet rays.

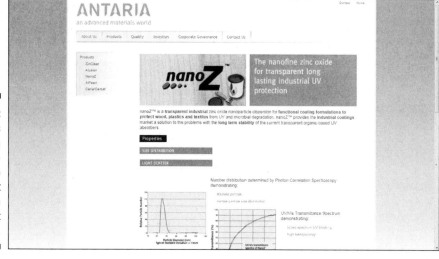

Figure 7-4: Anteria is producing nanoparticles to protect wood from ultraviolet rays.

For buildings with siding created from a porous material such as stone, a company called Nanoprotect (www.nanoprotect.co.uk) has created a coating called Nanostone, shown in Figure 7-5. This coating adds hydrophobic nanoparticles to the surface of porous material to give it waterproof properties that can help reduce buildup of mildew and even soot. They state that their product allows water vapor inside the structure to escape while blocking the entry of external water, reducing damage from trapped water.

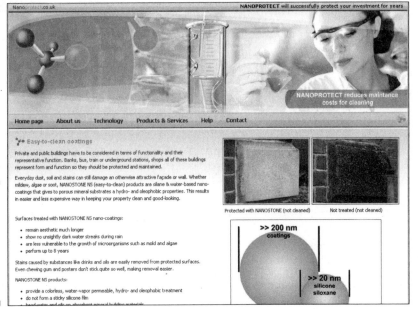

Figure 7-5: A comparison of how Nanostone protects stone surfaces.

A few companies are offering liquid products you simply paint on siding to provide both insulation and protection. Industrial NanTech, Inc., has an insulation product called Nansulate that is a nanocomposite liquid coating. When it dries, it provides exceptional insulation, as well as protection from corrosion, mold, and rust. If you apply Nansulate to interior or exterior walls, you can see a 20 to 40 percent savings in your energy usage. BASF also has a paint product called Col.9 that keeps your house siding cleaner and helps it last longer than traditional paint, as shown on their web site in Figure 7-6.

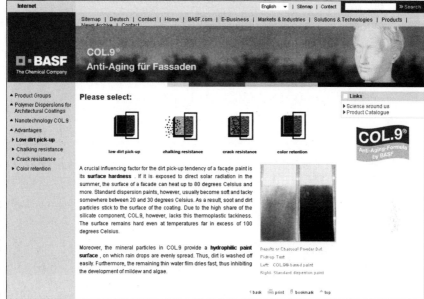

Figure 7-6:
The various attributes of Col.9 from BASF.

Keeping walls sterile

Because certain nanoparticles can act as photocatalysts, using energy from sunlight to cause reactions between organic molecules and the oxygen in water vapor, nanotechnology can be used to help break down bacteria on surfaces such as building walls.

A company call EcoActive Solutions makes OxiTitan (www.oxititan.com), a transparent antimicrobial coating that can be applied to walls and many other surfaces. OxiTitan, shown in Figure 7-7, uses zinc nanoparticles and nanocrystalline titanium oxide to break down bacteria. The coating is a photocatalyst, so some light is needed. Artificial light will do the trick, however, so the coating works indoors. Using OxiTitan in buildings such as your doctor's office, which is exposed to many germs on walls, countertops, and doorknobs, might prevent your trip to the doctor from making you sicker.

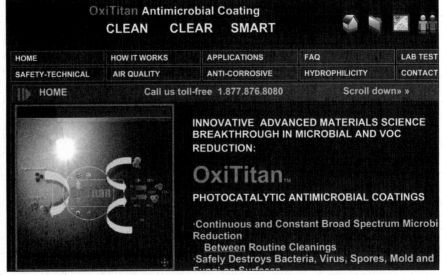

Figure 7-7: An illustration of OxiTitan showing how it breaks down bacteria.

Bioni (`http://www.bioni.de/`), whose web site is shown in Figure 7-8, makes several versions of paint containing nanoparticles of silver. Bioni-Hygienic is used in buildings such as hospitals and doctors' offices to help kill bacteria, and BioniNature is used in homes. Both products are designed to prevent the formation of mold and mildew in rooms such as bathrooms. Bioni's exterior paint contains nanoparticles of silver that help to stop the formation of moss and algae.

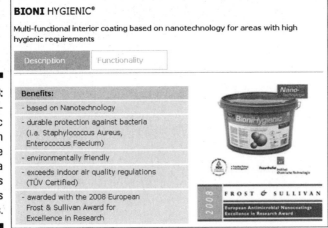

Figure 7-8: Bioni-Hygienic paint can help reduce bacteria in settings such as hospitals.

Tiling floors using a nanoleveling compound

When installing tile over an old floor, traditionally you had to use a leveling compound, then a board of concrete (called backer board), and then mortar to hold the tile in place. If mortar is applied over many types of surfaces, the tile can crack, which is why backer board is necessary. Nanotechnology can make the process simpler by incorporating nanopores, along with rubber granules, into the leveling compound to add some give to it. Because of that give, the backer board isn't required, making installation faster and less expensive.

BASF is using nanotechnology to improve the performance of leveling compounds used to prepare floors for laying tile in building renovations. PCI Nanosilent (see Figure 7-9) not only simplifies the tiling process but also reduces the sound that footsteps make when walking on the floor. BASF developed a curing process that produces nanopores in the dried leveling compound. The nanopores are then combined with rubber granules. The combination reduces the chance of the tile cracking if the old flooring expands at a different rate than the tile as a result of temperature changes.

Figure 7-9: PCI's Nanosilent could make your next tiling job easier.

Making concrete more durable with carbon nanotubes

One problem with concrete is that it has to be replaced after a few decades. Just walk the sidewalks of a town that hasn't bothered to do that and you'll see why.

By infusing concrete with carbon nanotubes, researchers at Northwestern University have discovered that they can extend the life of concrete. (You can read more about carbon nanotubes in Chapter 3.) When you look at concrete on the nanoscale, it's like a collection of balls packed together. When cement and water combine, a chemical reaction takes place that creates spaces between those balls called nanovoids. These nanovoids allow chips and cracks to begin. By using carbon nanotubes in concrete, you make the material much tougher and make it harder for those spaces to form, potentially expanding the material's life span from 20 years to 100 years.

This kind of concrete is more expensive, but it lasts about five times longer, so in the long run it's more cost effective. And creating less concrete to replace worn structures is also a boon to our environment because about 5 percent of the carbon dioxide we generate is produced from the manufacture of concrete.

Bright Ideas: Nanotechnology and Electronics

If you use a computer, you are probably using nanoelectronics. The integrated circuits that are the brains of the computer include nano-sized structures. For example, the smallest structure (called the minimum feature size) on the microprocessors that Intel is currently shipping is 32 nanometers in width. Microprocessors made with this process have as many as 995 million transistors packed on one computer chip.

Although the manufacturing process and minimum feature size vary from manufacturer to manufacturer, almost all microprocessors today have minimum feature sizes below 100 nm. And these microprocessors are not just in your computer; your car has several microprocessors running the electronics that monitor the engine, control your navigation system, and more. Do you have more nanoelectronic devices? Probably. If you have a cell phone, a television, a microwave, or an iPod or other music device, you have lots of nano in your life. (See Chapter 6 for a detailed look at how nanotechnology is used in electronic devices.)

Most electronic devices have some kind of display. Researchers at the University of Michigan have demonstrated that nanowires can be used as electrodes in organic light emitting diode (OLED) displays, thereby enabling manufacturers to build larger, more flexible OLED displays. This could affect the appearance and function of electronic devices.

Another way that nanotechnology will be used to change the way we think of electronic devices is by providing the capability to morph their shapes. Nokia Research Center (www.research.nokia.com/morph) is working on a mobile phone that you could bend to fit on your wrist like a bracelet, as shown in Figure 7-10. Nanoscale protein fibers are part of a mesh that provides a strong but flexible and stretchable material. In addition, nanostructured surfaces they call nanoflowers could help make the surface of a phone self-cleaning and water repellent, extending the life of the phone. A technology they call nanograss could produce solar power to run the phone. Finally, nanosensors as they would be used in Nokia's vision of the future could help us monitor levels of dangerous emissions or germs in our environment so we could navigate the world more safely and wisely.

Figure 7-10:
The Morph mobile phone could change the way you communicate.

Cleaning Up with Nano

After you build your house and populate it with electronic gadgets, you need to deal with the more mundane realities of housework. Cheer up! Nanotechnology is already being used in various cleaning products to make your life easier.

Companies are looking into using nanoparticles in soap that make it work better while producing less environmentally harmful byproducts. For example, EnviroSan Products (www.envirosan.com) offers a product called Solution 2000, and Nano Green Sciences (www.nanogreensciences.com/) produces a cleaning product called Nano Green. Both products contain organic nanoparticles, called micelles, which range in size from 1 to 4 nanometers in diameter. Several micelles bond to grease molecules, tying up all the atoms in the grease molecules that are attached to a surface, such as your countertop. After these micelles latch on, you can easily wipe away the grease molecules.

Some companies, such as AltimateEnviroCare Services (www.altimateenvirocare.com) and EcoActive Surfaces, are using titanium oxide nanoparticles as part of a film that uses the energy in light to kill bacteria on surfaces. Titanium oxide nanoparticles are called photocatalysts because of their capability to use energy in light to start the chemical reaction that kills the bacteria. OxiTitan, which we discuss in the "Keeping walls sterile" section, is a spray that coats a surface with zinc nanoparticles and titanium dioxide nanocrystals. This coating reacts with water in the air to break water down into oxygen and hydroxide ions. These ions then react with bacteria, viruses, volatile organic compounds, and mold, turning these organic molecules into carbon dioxide and less harmful organic molecules. Figure 7-11 shows the effect of this coating on a virus.

Some companies are using antibacterial materials that contain silver nanoparticles. Daido Special Steel Corporation has developed at spray called HGT Nano Silver Photocatalyst (www.hgt.com.hk/english/hgt.htm) that is a combination of silver nanoparticles and titanium dioxide nanoparticles. This product performs when light is available, with the silver nanoparticles enhancing the photocatalytic performance of the titanium nanoparticles. However, because silver nanoparticles kill bacteria even when light is not available, the treated surface will have antibacterial properties even in the dark. At this time, the product is available only in Japan. Figure 7-12 illustrates the process of photocatalytic nanoparticles breaking down bacteria into carbon dioxide and water vapor.

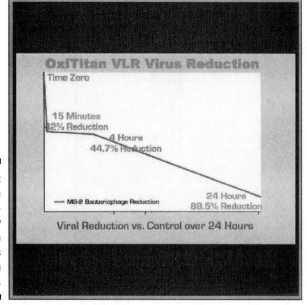

Figure 7-11:
A graph showing how OxiTitan reduces bacteria on surfaces.

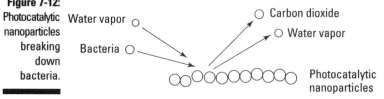

Figure 7-12:
Photocatalytic nanoparticles breaking down bacteria.

Note that there is interest in regulating the use of silver nanoparticles. The concern is that when silver nanoparticles are released, they may harm useful bacteria in groundwater, such as lakes.

Applying Nanotechnology to Cars

The automotive industry is hot on the tail of nanotechnology. Nano will help them power vehicles more cost effectively, protect the surface of cars and windows with nano paint and nano coatings, and make stronger tires. You'll see these and more advances in the cars of the future, courtesy of nanotechnology.

Charging up your car with the sun

One of the companies using nanotechnology to develop low-cost solar cells, as discussed in Chapter 10, has a way to make the paint on the outside of your car into one big solar cell.

Global Photonic Energy Corporation (www.globalphotonic.com/), whose web site is shown in Figure 7-13, says that you can use a spray painting method to paint their solar cells onto a car or other places that use spray paint, such as cell phone cases. The paint can come in virtually any color. They want to license their PowerPaint solar cells to manufacturers, but as of this writing, they haven't announced which car manufacturer will be the first to use it.

Figure 7-13:
PowerPaint
could be
one product
that can
bring solar
cells to the
outside of
your car.

Powering electric and hybrid cars

Electric and hybrid cars are becoming more popular given the cost of a tank of gas. Nanotechnology can improve batteries by increasing the surface area of the electrodes, which allows the battery to store more charge.

Work by nanotech battery companies such as Altair Nanotechnologies and A123 Systems (www.a123systems.com) could improve the performance of lithium-ion batteries and may make electric cars even more appealing. Fisker Automotive says that its Karma plug-in hybrid car using A123's Nanophosphate battery (see Figure 7-14) should be available in the spring of 2011 and will have an average mileage of 67 mpg. Pretty good mileage for a sports car!

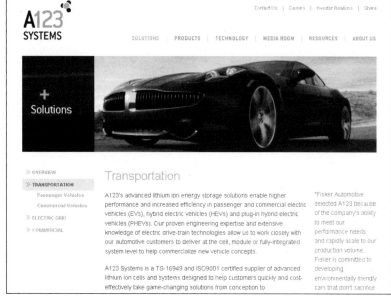

Figure 7-14:
Nano-
phosphate
batteries
are used in
Fisker Auto-
motive's
plug-in
hybrid elec-
tric car.

Grasping the potential of hydrogen fuel cells

You may have heard talk about cars powered by hydrogen fuel cells replacing gasoline-powered cars, but don't hold your breath. The major obstacles to widespread use of hydrogen fuel cell–powered cars in the next few years are the lack of a network of hydrogen fuel stations, the high cost of hydrogen fuel cells, and the need for lightweight and safe hydrogen fuel tanks.

Nanotechnology may help with the cost of fuel cells by using nanoparticle-based catalysts to reduce the amount of platinum required. In addition, because hydrogen bonds very strongly to carbon, we can make lighter, safer fuel tanks by using graphene, which has the largest amount of carbon atoms per weight on its surface of any material.

See Chapter 10 for a more detailed discussion of how nanotechnology is being used in the production of hydrogen fuel cells.

These hurdles mean that hydrogen fuel cell–powered cars won't be your way around high gas prices next summer. The Department of Energy's Hydrogen Program estimates that the start of mass market usage of hydrogen fuel cell cars won't happen until 2020.

Keeping that paint job shiny

Everyone who has ever owned a new car knows the heartbreak of that first big scratch in the paint job. Nanotechnology could help you out because several companies are producing scratch-resistant products that could be built into your car's paint job or applied aftermarket. These products use nanoparticles that form chemical bonds to the car surface, creating a longer-lasting coating.

One company, PPG, produces a scratch-resistant clearcoat called CeramiClear that helps a car's paint stay glossy longer. It contains nanoparticles that bead on the surface and produce the longer-lasting shine. In addition to the gloss, CeramiClear protects the paint from chipping by producing a surface similar to silica (glass).

Mercedes uses CeramiClear on some of their models. If you're thinking of picking up a new Mercedes, you should know that the paint code for models using CeramiClear starts with a C.

Some aftermarket products containing nanoparticles may also be useful for protecting your car's paint job. A company called Matrix Micro Coatings (http://matrixmicrocoatings.usalocalmall.com), whose web site is shown in Figure 7-15, has a product called NanoGloss that uses hydrophobic nanoparticles to repel water and make cars easier to clean.

Another company, Nanolex (www.nanolex.de/en), whose web site is shown in Figure 7-16, also makes a nanoparticle-based sealant to protect your paint job. According to their web site, "Surfaces sealed with Nanolex repel water, oil, and dirt, have antistatic characteristics, and protect against chemical and biological damage. Water, oil and dirt can be removed easily." This information may be worth knowing if you're concerned about chemical or biological attacks in your neighborhood.

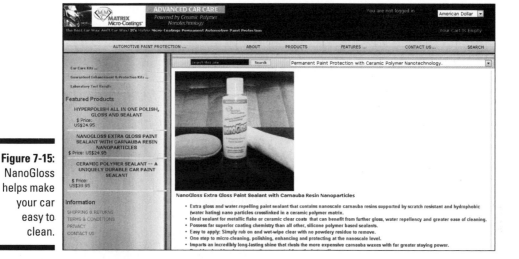

Figure 7-15:
NanoGloss
helps make
your car
easy to
clean.

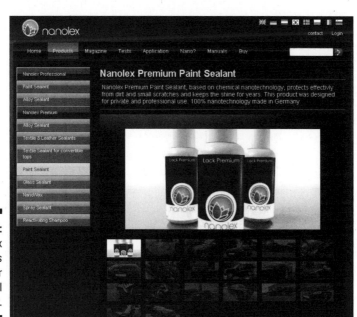

Figure 7-16:
Nanolex
protects
your car
against oil
and dirt.

Treating car windows

Are you sick of those little bug splats on your windshield? We are, too. That's why we were glad to stumble across NanoSafeguard Auto Glass Treatment, shown in Figure 7-17 (www.nanowindshield.com/). Spread this nano-based coating on window glass, and it allows only minimal contact with the things it encounters, from bugs and dirt to ice. This may be bad news for the car wash and de-icer product industries but good news for consumers like you.

Figure 7-17:
This coating can give you a clear view without windshield wipers even in a downpour.

But this product goes beyond saving you from dirty windshields. In time, it's possible that windshield wipers will become obsolete because NanoSafeguard Auto Glass Treatment would repel most of the water and oils that assault your windshield even in rainy weather, and the wind whipping around your car would blow away the rest. This type of coating uses nanoparticles containing atoms on one end that form chemical bonds to the glass, and atoms on the other end that are hydrophobic (that is, they repel water).

A company called Nanofilm (www.nanofilmtechnology.com/) makes Clarity Defender Automotive Windshield Treatment. This thin film can be used on both mirrors and glass in your car. It produces a nanoscale barrier to water, snow, ice, as well as those nasty bug splats. One of the claims of this company is that the film helps to increase a driver's visibility, improving response times to road dangers.

Checking the tires

Some say a car is only as good as its tires. They grip the road and spin you along your way. So what is nanotechnology doing to improve tires? Nanoparticles can be used to bond the various substances in the tread, strengthening the material.

A company called Yokohama (`www.yokohamatire.com/`), whose web site is shown in Figure 7-18, has developed a tire called ADVAN Sport that uses a nanotechnology-enabled tread compound to help the tires grip the road. This tire is intended to provide higher performance than most of us will ever need.

Figure 7-18: Tires that grip better from Yokohama.

Another company called Lanxess (`www.lanxess.com`) has developed a nanoparticle to make tires last longer, have a better grip, and reduce resistance (which can save fuel). Nanoprene particles of rubber compound in the tire provide anchor points to attach to a silica filler.

Making cars lighter weight

Various car manufacturers are researching the possibility of using nanocomposites to create strong, lightweight materials that can be used in a car body to reduce the weight of the car. These materials combine nanoparticles with polymers to create lightweight materials as strong as steel.

For example, Ford is helping fund research at the Advanced Materials Lab at Northwestern University (`www.mech.northwestern.edu/fac/brinson/new/research/`), whose web site is shown in Figure 7-19. The material they are developing is scratch-resistant, lightweight, and rust-proof. It also makes car bodies stronger and lighter weight, which translates into a longer car life span and savings at the gas pump, respectively.

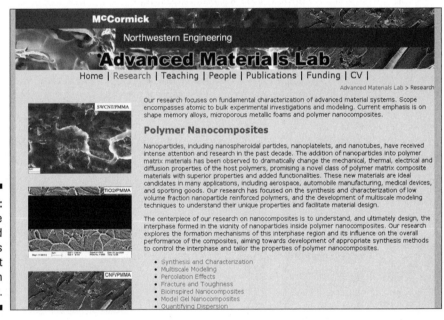

Figure 7-19:
The
Advanced
Materials
Lab at
Northwestern
University.

Chapter 8

When Nano Gets Personal

*N*ano has already found its way into lots of products you use every day, from clothing to tennis racquets. In fact, if you strolled around your home you'd probably find dozens of products manufactured using some kind of nanotechnology.

In this chapter, we cover sporting goods that are stronger, bounce higher, or go further because of their nano properties; fabrics that shed dirt and water or keep you warmer; food and food packaging that has been enhanced with nano; and cosmetics such as sunscreens and makeup that help you look younger or protect your skin with nano properties. Consider this a minisampling of some products already on the market and others under development, while remembering that new products and techniques come out all the time.

To keep up with changes in this area, visit this page on the companion web site: www.understandingnano.com/nanotechnology-consumer-products.html.

Seeking Sleeker Sporting Goods

Nanotechnology and sporting goods are a logical match, when you consider that nano can make things lighter, bouncier, stronger, or more flexible. From stronger tennis racquets to more slippery skis, nano is a player.

Making tennis balls that bounce longer

The hiss when you open a new can of tennis balls is because the can is full of pressurized air. After the can is open, the pressurized air inside the tennis balls will slowly leak out until the pressure inside the ball equals the air pressure and the balls lose their bounce. If you've ever hit the tennis courts, you know that a sluggish ball is a real downer. Wilson has introduced a Double Core technology in their tennis balls that uses a nanocomposite coating that makes the balls bounce twice as long as other balls. The nanocomposite coating used in these balls is a mix of rubber and nanoclay particles that provides a gas barrier that slows down the loss of pressurized air from the tennis ball. This nanocomposite material allows you to use the ball longer at full performance.

The tennis balls are currently manufactured by InMat LLC (www.inmat.com), whose web site is shown in Figure 8-1.

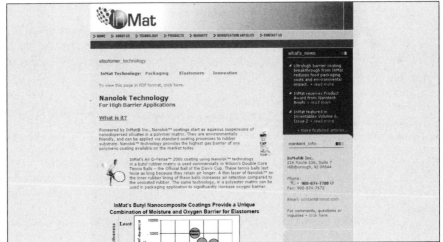

Figure 8-1:
Tennis balls
from InMat.

Producing lightweight and powerful racquets

If you've used a tennis racquet made by Wilson Sporting Goods that was manufactured after 2004, chances are you are using nanotechnology every time you hit the ball. In 2004, Wilson (www.wilson.com) started selling nCode tennis racquets, which contain silicon dioxide nanoparticles that fill gaps between carbon fibers in the racquet strings to add strength.

Then in 2007, Wilson started selling tennis racquets with the [K] Factor designation, in which the silicon dioxide nanoparticles not only fill the gaps between the carbon fibers but are also bonded to the carbon fibers, which produces a better feel and a more stable racquet. And in 2010, Wilson introduced the BLX line of racquets (shown in Figure 8-2), which replaces the carbon fibers with basalt fibers while keeping the silicon nanoparticles to bond the basalt fibers together. Wilson states that this change helps to reduce vibration in the racquet and give players better control. Basalt fibers also cost significantly less than carbon fibers, so Wilson may be reducing its cost, as well as improving the performance of their racquets.

Figure 8-2:
BLX rac-
quets from
Wilson.

Wilson also uses the BLX technology in racquets for badminton, squash, racquetball, and platform tennis.

Buckyballs and nanotubes are very strong, as we discuss in Chapter 3, making them ideal for adding strength to composite materials without adding significant weight. However, because buckyballs and nanotubes are more expensive than many other types of nanoparticles, most manufacturers have not adopted them. An exception is a company called Yonex (www.yonex-usa.com), which makes racquets using carbon nanotubes and functionalized buckyballs that bind to carbon fibers in badminton and tennis racquets to optimize power and flex while keeping the racquets lightweight.

Going golfing

For those interested in golfing, the Holy Grail of a golf club that makes golf balls go faster and farther is something worth seeking. A few companies are working towards that goal using nanotechnology.

Yonex has a Nanospeed line of golf clubs, shown in Figure 8-3, that uses carbon nanotubes, which they say helps to make the club heads stronger and to transfer energy through the shaft more efficiently, which translates into golf balls that travel faster. Another company using nano for lightweight and efficient drivers is Maruman (www.maruman.co.jp) with its Exim Nano line of clubs.

Figure 8-3:
Nanospeed golf clubs that contain carbon nanotubes.

. . . and more

We uncovered a few more applications of nanotechnology in the sports arena, including the following:

- ✔ Cycling: Easton Cycling (www.eastonbellsports.com) uses carbon nanotubes in a resin to reinforce the carbon fibers in their handlebars and cranks, which are shown in Figure 8-4. This process increases the stiffness of the handlebars and cranks while keeping them lightweight.

- ✔ Skiing: Holmenkol (www.holmenkol.com) has a ski wax called nano-CFC that uses nanocomposites. Their web site is shown in Figure 8-5. Using this wax increases what they call "abrasion resistance" and water repellency, which make the skis glide better and faster on snow.

- ✔ Fishing: The St. Croix Rod Company (www.stcroixrods.com), whose rods are shown in Figure 8-6, is using nanoparticles of silica in a resin they call NSi (Nano Silica). NSi fills the space between carbon fibers in their fishing rods, which they say results in a 30 percent stronger rod that's still lightweight and flexible.

Figure 8-4:
Easton
Bell Sports
builds nano
bikes.

Figure 8-5:
Holmenkol's
ski wax
adds glide
to skis.

✔ Kayaking: The Norwegian research company ReTurn AS (www.re-turn.
no/) has developed an epoxy gelcoat that has been modified with carbon
nanotubes to treat the outer skin of kayaks. This coating makes the kayaks
more resistant to abrasion and cracks.

Figure 8-6:
St. Croix Rod Company fishing rods.

✔ Archery: Easton Archery (www.eastonarchery.com) makes arrows with a resin containing carbon nanotubes in their N-FUSED CARBON AXIS arrows. This resin produces lightweight arrows that are stronger and control vibration better. The company's web site is shown in Figure 8-7.

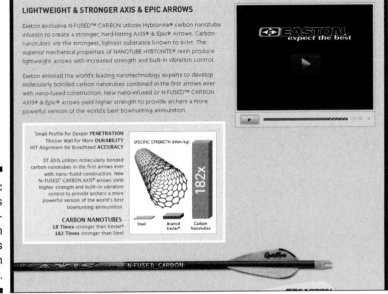

Figure 8-7:
Arrows containing carbon nanotubes from Easton Archery.

Helping Fabrics Do More

As an improvement on past techniques, making fabric with nano-sized particles allows manufacturers to improve fabric properties without a significant increase in weight, thickness, or stiffness. Here's a sampling of what's happening today in the area of fabrics and clothing with a nano spin.

Making fabric water and stain resistant

Manufacturers are using a few methods for making clothing water and stain resistant. Two companies turn to the lotus leaf for inspiration, while another takes its cue from peaches. Both ideas are likely to be pretty darn fruitful (pun intended).

Contemplating the lotus leaf

You have probably noticed the clothing manufacturers who advertise their clothes as water and stain resistant; many of them are using nanotechnology to achieve these characteristics. These companies have essentially found a nano way to copy the way plants, such as the lotus leaf, shed water.

The top of a lotus leaf has a rough surface because of little waxy spikes, which cause moisture to bead up and slide off the leaf. A company called Schoeller Technologies has a fabric treatment called NanoSphere that adds nanoparticles to the surface of fabric, allowing clothing to shed water just like a lotus leaf does. The hilly surface of NanoSphere results in less area with which dirt or water can make contact. The NanoSphere web site (www.nanosphere.ch/) is shown in Figure 8-8.

Figure 8-8: NanoSphere fabric products.

Another company, BASF, has created a product called Mincor TX TT. Their process is similar to Schoeller Technology's in that they too use the so-called lotus effect. BASF's trick is to pack billions of nanoparticles onto fabric so closely that dust and dirt can't get through and attach themselves to the fabric. Dirt simply stays in a layer of air above the fabric and washes off easily. Mincor is used in awnings, umbrellas, and tents.

Repelling water and stains by studying peaches

A company called Nano-tex (www.nanotex.com), whose web site is shown in Figure 8-9, modeled their method to make clothing and upholstery fabrics water and stain resistant after the way fuzz on a peach repels water. Nano-tex is using what they call nano-whiskers. Similar to the fuzz on a peach, these whiskers are tiny hair-like projections that cause liquid to bead up and roll off the surface of fabric.

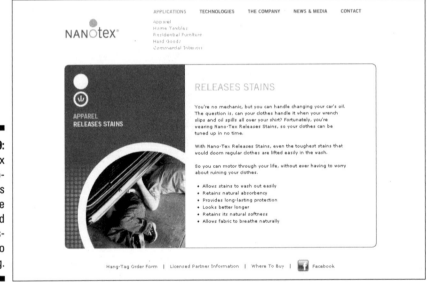

Figure 8-9:
Nano-tex uses nano-whiskers to provide water and stain resistance to clothing.

These whiskers are aligned along spines using what the company refers to as molecular hooks. This system of whiskers and hooks make the fabric more durable but don't make it less breathable (meaning that the material still lets air in and out so you don't feel like you're in a sauna). Nano-tex claims that, where other treatments reduce breathability and wear off over time, their treatment lasts longer.

Making fabric that produces or stores electricity

Put aside that image of banks of solar cells on a building roof for the moment and consider the possibility of tiny nanosized solar cells you can carry around on the surface of your backpack. That possibility of producing energy, along with the potential for fabric to store energy, has become a reality. The products already being produced or developed using such fabrics allow you to charge your electronic devices using your tent, awning, or even the clothing you wear.

Drawing energy with solar cells

Wearable, nano-enabled solar cells are available in stores today in the form of backpacks made by a company called Traveler's Choice. This product contains Konarka Technologies' Power Plastic, a flexible solar cell that uses organic molecules with nanoparticles embedded in a plastic substrate to improve the efficiency of the solar cells.

If you walk around with a Traveler's Choice backpack on a sunny day, it can charge your cell phone or extend the battery life of any electronic device. The same technology is used on a briefcase, such as the one shown in Figure 8-10.

Figure 8-10:
Solar-powered briefcase.

Traveler's Choice

KONARKA®

Case Study

Traveler's Choice Solar Bag

The Challenge: Bring new excitement and solar capabilities to traditional backpacks and travel bags.

The Solution: Design stylish solar bags that power handheld devices from the sun—and on the go.

Traveler's Choice, one of the world leaders in fine travelware, selected Konarka Technologies Power Plastic®

Travelware leader Traveler's Choice prides itself on being ahead of the curve with innovations and enhancements for its extensive line of practical, attractive, and advanced travel goods—from suitcases to briefcases to duffels and more. Encompassing Traveler's Choice, Coleman,

the mass market. With its new bags, Traveler's Choice is introducing an innovative, next-generation of travel bags—all fueled by Power Plastic.

Simple handling and fabrication
Fabrication was simple, thanks to the

Konarka's current research focuses on using Power Fiber to seamlessly weave photovoltaic material (the material which makes up solar cells) into fabric. Rather than being layered onto a plastic substrate, Power Fiber is made by coating the primary electrode with consecutive layers of active material made from nanoparticles, the transparent electrode, and the transparent substrate. Fabrics can have the same look and feel as regular fabrics, while being able to produce power.

Other companies are also working on nano-enabled solar cell material that can be woven into fabric. For example, Solarmer Energy (www.solarmer.com), whose web site is shown in Figure 8-11, says it will have such a product available in 2011. This company's plastic solar cells are intended for use in smart fabrics. The cells can be "printed" on the fabric for products such as jackets, tents, awnings, and suitcases using nanoparticle ink.

Figure 8-11:
Solar cell fabric manufacture Solarmer's web site.

Global Photonic Energy Corporation (www.globalphotonic.com) has created a technology that it calls SM-OPV FlexPower to create solar cells with high levels of performance. These flexible, organic cells can be placed in plastic substrates or woven into fabric in the form of power-generating threads.

Using fibers to store energy

Batteries store energy. Your standard battery contains a metallic foil that is covered with a substance called an electrolyte and then coiled inside the battery. At Stanford, researchers have now produced batteries using fabric. The fabric is soaked in ink containing nanoparticles. One piece of fabric is soaked in nanoparticles that make it act like an anode, and another piece to which it's attached is soaked in different nanoparticles to make it act like a cathode. (The exchange of ions between an anode and a cathode causes a battery to release or store electrons.) This fabric, which the researchers have named eTextiles, is much less costly to produce than the metallic foil used in traditional batteries.

If you vary the ingredients in the ink into which the fabric gets dipped, you can create a supercapacitor, which is another device used to store energy. The ink used to make the batteries contains oxide nanoparticles, and the ink used to make the supercapacitors contains single-walled carbon nanotubes. Because textiles soak up the conductive ink so well, they make very efficient storage devices. They can hold about three times as much energy as your mobile phone battery and are stronger and more durable.

Being able to store power on your person could make it possible to have your clothes monitor your health or display messages. Perhaps the latter would be handy if you want to flirt with that person across a crowded room?

Clothing that keeps you warmer

Clothes don't just make the man or woman, they also keep you warm or cool. A material called aerogel has been developed that can help provide insulation in clothing. Aerogel is composed of nanoparticles interspersed with nanopores filled with air, which makes it one of the best thermal insulators in existence. We discuss the discovery of aerogels in more detail in Chapter 12.

Companies such as Aspen Aerogels (www.aerogel.com) and Cabot (www.cabotcorp.com) have introduced materials using aerogel for outdoor clothing that insulates against extreme temperatures. Nanoscale air pockets in the material keep cold and heat out without adding bulk or weight to the clothing.

Climbers testing out jackets using Aspen Aerogels, whose web site is shown in Figure 8-12, reported that they were comfortable on the top of Mount Blanc at temperatures of –20 to –25 degrees centigrade and in winds of 70 to 80 km per hour.

Figure 8-12:
Products
from Aspen
Aerogels
are used in
clothing that
keeps folks
warmer.

Shoe insoles and camping pads are also available using Aspen Aerogels, but a climber who reached the top of Mount Everest wearing Aspen Aerogels insoles reported that her feet were too warm. The companies that were manufacturing jackets a few years ago are not showing them in their product line at this time, possibly because they were also too warm.

Hanes brands may have solved that problem. The company is currently working on incorporating Zero-Loft Aerogel from Aspen Aerogels into a jacket that was spotted at a trade show in early 2011. Their jacket and accompanying pants will be called the Champion Supersuit and will offer the thinnest outerwear for extreme weather ever introduced.

The aerogel in this outerwear is supposed to be more than 99 percent air trapped in tiny chambers that will keep you about four times as warm as that goose down comforter on your bed. This type of thin insulating clothing (only a few millimeters thick) may soon replace those puffy quilted coats seen in chilly climates and on mountain treks, for which the fashion-conscious among us are grateful.

Cabot Corporation (www.cabot-corp.com), whose web site is shown in Figure 8-13, has also developed an aerogel-based fabric called Nanogel Thermal Wrap to insulate jackets, gloves, boots, and more.

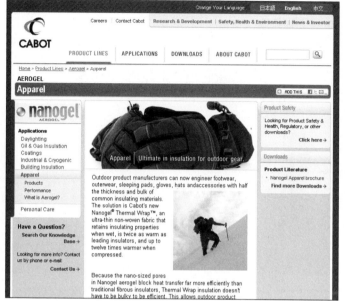

Figure 8-13:
Cabot
Corporation
produces
an aerogel-
based fabric
that insu-
lates.

Clothing that makes you smell better

Want to smell like a rose even when running in your tracksuit? Soon you may be able to.

Microorganisms cause most bad odors. The good news is that microorganisms in the form of bacteria can't live around silver ions. A new product called SmartSilver from NanoHorizons (www.nanohorizons.com) uses silver nanoparticles to kill odors in clothing, medical materials, and household goods. This technology generates silver ions from the surface of silver nanoparticles, making it a more effective odor killer. The silver nanoparticles bond at the molecular level to the fabric in this process, which translates into odor protection that literally never wears out (unlike your average room freshener).

SmartSilver is used in socks made by Wigwam (www.wigwam.com), whose web site is shown in Figure 8-14.

Figure 8-14:
Keeping socks odor free with SmartSilver.

SmartSilver is also available in shoes made by Zoot (www.zootsports.com). Their linings use antimicrobial SmartSilver to reduce moisture, odor, and hot spots. These shoes, shown in Figure 8-15, are perfect for runners.

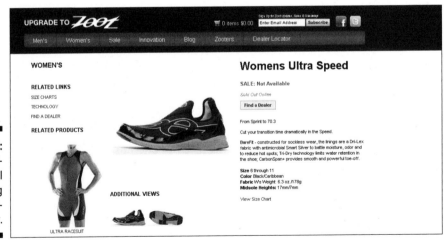

Figure 8-15:
Zoot's anti-microbial shoes using nanotechnology.

The antimicrobial properties of SmartSilver seem to be a natural for clothing used by medical professionals. You can find scrubs using SmartSilver on the Nursing Scrubs Shop web site (www.nursingscrubsshop.com).

Another manufacturer named Odegon Technologies offers teabag-sized bags to help freshen your clothing. The bags (which they call tags) contain nano-porous material that rounds up and stores molecules that cause odors.

Creating fabrics with a unique fit

When you talk about skin-tight clothing, you're using an apt metaphor. Cells and tissues come together in a structure of protein fibers that allow our skin to fit snugly around the surface of our body while retaining some elasticity.

Bioengineers at the Wyss Institute for Biologically Inspired Engineering (http://wyss.harvard.edu), whose web site is shown in Figure 8-16, and the School of Engineering and Applied Sciences (SEAS) at Harvard University have copied the properties of skin in the lab. Their new technology allows them to make fabrics that are only a nanometer thick. Thanks to nanotechnology, these fabrics are strong but also flexible.

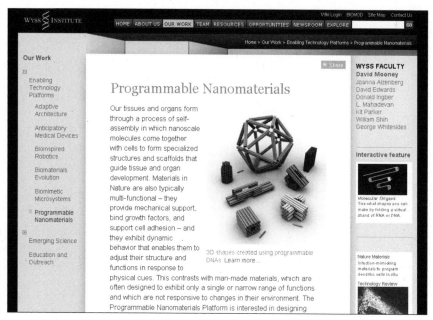

Figure 8-16: Making clothing act like skin.

Says Adam W. Feinberg of Harvard, "To date it has been very difficult to replicate this extracellular matrix using manmade materials. But we thought if cells can build this matrix at the surface of their membranes, maybe we can build it ourselves on a surface too. We were thrilled to see that we could."

The ability to change the kind of protein they use in this material allows researchers to adjust thread count and other properties, making possible textiles that can be stretched as much as 1500 percent from their original size. This ability could make these fabrics useful in form-fitting bandages or clothing that adjusts its size depending on who is wearing it (or right after Thanksgiving dinner).

The army is developing a nanobattlesuit that could be bullet-resistant, as thin as spandex, and even contain health monitors and communications equipment. These jumpsuit-style outfits might even be able to react to and stop biological and chemical attacks. All the pieces would be integrated into an efficient lightweight suit. We discuss this battlesuit in more detail in Chapter 12.

Just Desserts: Optimizing Food

Nanotechnology is having an effect on several aspects of the food industry, from how food is grown to how it is packaged. Researchers are developing nanomaterials that will make a difference not only in the taste of food but also in food safety and the health benefits that food can deliver.

How is the food industry dealing with nanotechnology?

While researching the food industry and nano for this book, we came across an interesting phenomenon: Most major food companies are not performing nanotechnology research or funding nanotechnology research into improving food. Because we knew there were some very logical applications of nano to food science, we contacted a few professors of food science to get the inside scoop.

One reason that large food companies appear to be shying away from nano is that they have concerns about potential legislation. For example, one piece of legislation being worked on in the European Parliament may limit the sale of foods produced with nanotechnology. Writing such legislation or regulations will be challenging because it's difficult to restrict the presence of nanoparticles in food: Food can contain naturally occurring particles between one and one hundred nanometers in diameter, a standard definition of nanoparticles. One option is that they might restrict the presence of only engineered nanoparticles and not naturally occurring ones. Still, the confusion and uncertainty surrounding these regulations is making the industry cautious, especially after run-ins with the general public's perception of genetically engineered foods a few years ago.

One important area for food companies and regulators to deal with is how to keep food from becoming contaminated with bacteria such as Salmonella or E. coli. Nano methods might be developed to stop pathogens from growing in food or track contaminated food to its source. In talking with Professor John D. Flores, head of the Food Science department at Pennsylvania State University, he expressed concern that slower research into the use of nanotechnology in food might also slow the development of methods to make our food supply safer.

That said, areas of research and development are going on in the area of food science primarily at universities funded by government money. In this section we cover a few of them.

Nanomaterials in food packaging

The use of nanomaterials in food packaging is already a reality. For example, bottles made with nanocomposites minimize the leakage of carbon dioxide out of the bottle, increasing the shelf life of carbonated beverages without using heavier glass bottles or more expensive cans.

Honeywell (http://honeywell.com) produces a resin called Aegis (see Figure 8-17) that is used to make plastic beer bottles. Clay particles, about a nanometer thick, are dispersed throughout the plastic resin so that there is no straight path for carbon dioxide molecules to use to flow out of the bottle.

Another example of nanoclays used in food containers is from a company called InMat (www.inmat.com). They use nanoclay platelets about a nanometer thick in a solution called Nanolok (see Figure 8-18) that can be applied to plastic films to prevent oxygen and water from penetrating the container, thereby increasing the shelf life of food. Nanolok is used mainly with dry food such as snacks, nuts, and coffee.

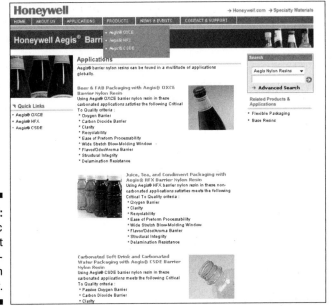

Figure 8-17:
Plastic
bottles that
keep bever-
ages fresh
longer.

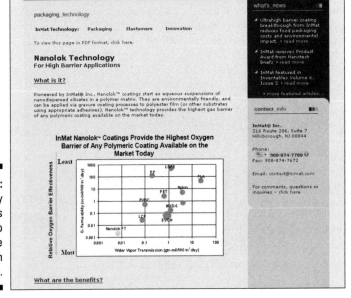

Figure 8-18:
Nanoclay
containers
to keep
moisture
away from
dry foods.

Several companies have produced food storage bins that include silver nanoparticles embedded in the plastic. The silver nanoparticles kill bacteria from any food previously stored in the bins, minimizing harmful bacteria.

Other nano food packaging products are currently under development. For example, research is being conducted on nanosensors in plastic packaging that can detect gases given off by food when it spoils and then change the packaging color. Researchers at various universities have tackled this idea, but no product has yet hit the shelves of your local grocery store.

Sensing food impurities

One of the big challenges in food science is to detect the kind of impurities that can spread contagions such as salmonella.

Companies are developing nanosensors that can detect bacteria and other contaminates such as salmonella and E. coli on the surface of food at a packaging plant. These sensors will allow for frequent testing that costs much less than sending samples to a lab for analysis. This point-of-packaging testing, if conducted properly, has the potential to dramatically reduce the chance of contaminated food reaching grocery store shelves.

A company called Nano Science Diagnostics (`http://nanoscience diagnostics.com`), whose web site is shown in Figure 8-19, has developed a diagnostics system that can test for bacteria on the spot in 15 minutes.

Figure 8-19:
Nano Science Diagnostics nano-enabled sensors detect food impurities.

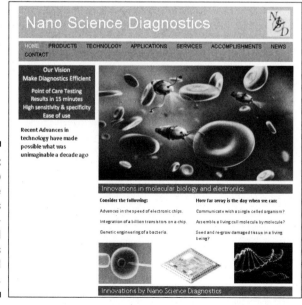

Folks are also developing nanosensors to detect lingering pesticides on fruit and vegetables. Researchers are working on an electronic nose that would be more sensitive than a dog's nose and be able to detect a few molecules of the pesticides in the air above vegetables. Although this would be useful at a packing plant, we're anxiously waiting for the handheld version so we can check out the apples and grapes in our local grocery store!

Changing food characteristics with nanomaterials

Nanoparticles are being used to deliver vitamins or other nutrients in food and beverages without affecting their taste or appearance. These nanoparticles encapsulate the nutrients and carry them through the stomach into the bloodstream. In some cases, this delivery method also allows a higher percentage of the nutrients to be absorbed by your body. When encapsulated by the nanoparticles, some nutrients that would otherwise be lost in your stomach are instead absorbed.

Researchers are also looking into developing nanocapsules containing nutrients that are released when nanosensors detect a deficiency in your body. Basically this research could result in a super-vitamin storage system in your body that gives you just what you need, when you need it.

Scientists are also developing nanomaterials that could change the taste, color, and texture of foods you eat. So-called interactive foods are being developed that would allow you to choose the flavor and color of your food. Want lemon-lime soda instead of root beer? Just program your beverage! The idea is that nanocapsules that contain flavor or color enhancers sit in the food waiting until a hungry or thirsty consumer triggers them. The method hasn't been published, so it will be interesting to see how this particular trick is accomplished.

Britain's Daily Mail Online described one such experiment that uses nanoparticles, comparing it to a stick of gum in the Willy Wonka story that produced the flavors of a full roast beef dinner. Apparently the Institute of Food Research (IFR) has stated that current technology could help us program and release certain flavors. Each flavor in the meal is separated by a gelatin layer that has no flavor at all. Researchers at Harvard are working on nanostructures they call colliodosomes that can capture the flavor of various ingredients. These flavors would be released based on enzymes in saliva or the time spent chewing the gum or other substance.

Unfortunately, at this point, such a product would only give you a taste, but not make you feel full nor provide any nutritional value. (But on the plus side, the product would have no calories!)

Finally, various reports have said that researches are developing nanoparticle emulsions for use in ice cream to improve its texture and uniformity.

The estimates of how many food products or packaging materials have nano components vary widely, to some degree because of differing definitions of nanotechnology. As a result of this and regulatory uncertainty, food companies generally are stating that nano components have not been added to their product.

Using nanotechnology to grow food

One big concern in growing food is the use of pesticides. Put simply, you have to kill the bugs that damage crops without killing the people who eat the crops. Researchers are working on pesticides encapsulated in nanoparticles that release pesticide only in an insect's stomach, minimizing the contamination of plants themselves and the risk to people.

Another development in the area of agriculture is a network of nanosensors and dispensers that can be deployed throughout a food crop. The sensors recognize when a plant needs nutrients or water, before a farmer can see any sign that the plant is suffering. The dispensers then release fertilizer, nutrients, or water as needed, optimizing the growth of each plant in the field one by one. To make this a reality, you need to be able to build millions of inexpensive sensors.

Researchers have been looking at this problem for several years. Hewlett Packard Labs is working on a worldwide solution. The HP program is called Central Nervous System for the Earth (CeNSE). Their goal is to create a global network of sensors that is inexpensive, resistant to damage, and incredibly sensitive. The first sensor that will be tested is a motion and vibration detector embedded in a silicon chip built with three layers. If the chip moves, the movement of the wafer suspended in the middle can be measured. The next logical candidates for sensors are those to detect light, temperature, barometric pressure, and humidity. HP's experience with inkjet printer cartridges may prove helpful as they package technology into an integrated unit. Their memristor technology could also provide logic and memory in a small sensor package that could be powered with a very small amount of energy.

The sensitivity of these sensors is about 1,000 times more than the devices used in a Wii or car airbag system.

One big challenge is to reduce the size of these sensors. Also, to truly be able to provide a global system of sensors, their cost must be miniscule because trillions would be needed.

Today the state of this technology is not ready for prime time because the detectors are expensive and turn in many false alarms. When these problems are solved, such sensors could help in food production in many ways. Eventually, sensors could be added to mobile phones. Then food science workers out in the field could wave their cell phones over a head of lettuce and the phone could sense the presence of salmonella or dangerous levels of pesticides.

Skin Care That Keeps You Young

Eventually, nanotechnology may help us reverse aging at a cellular level (see Chapter 9 for more about this). Until that day comes, we'll have to be content with the ways that nanotechnology is being used in cosmetics to keep our skin more youthful and provide protection from harmful sunlight.

Providing vitamins and nutrients for that youthful glow

One company is using nanotechnology to deliver vitamins and other nutrients to your skin cells. Marie Louise Cosmetics (www.marielouisecosmetics.com) uses a nanoemulsion to get these goodies to penetrate your skin. The nanoparticles in the three-layer emulsion are approximately 40 nm to 100 nm in diameter. The company states that the nanoparticles in the emulsion are designed to penetrate farther into the skin than emulsions that use lager particles and to release vitamins from the outer layers of the nanoparticle as it passes through the outer layers of the skin.

Figure 8-20 shows one of their products that contains nanoparticles on the company web site.

Some ethical concerns about the use of nanoparticles in skin care products exist because there is little regulation and it's not known if there could be long-term side effects. See Chapter 13 for more about this issue.

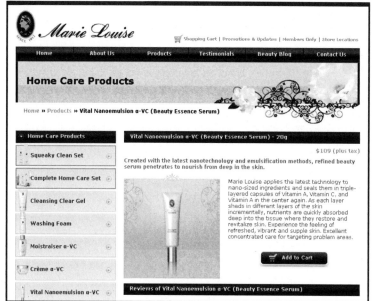

Figure 8-20:
Cosmetics
to keep you
nanoyoung.

The Fountain of Youth:
Anti-aging products

Although we can't yet reverse aging, one company is using nanotechnology to do some basic cellular repair in skin. Lifeline Skin Care (www.lifeline skincare.com) has a line of products that they claim can rejuvenate skin cells. They use stem cells (the nonembryonic kind), which is a type of cell that has the capability to stimulate the rejuvenation of other cells. The stem cells produce proteins that direct cells to build new cells to repair damaged skin tissue. The company encapsulates the proteins in nanoparticles.

When you spread the serum containing the nanoparticles on your skin, they open, delivering proteins directly to your skin. Figure 8-21 shows an illustration of the nanoparticle used to deliver the proteins from Lifeline's web site.

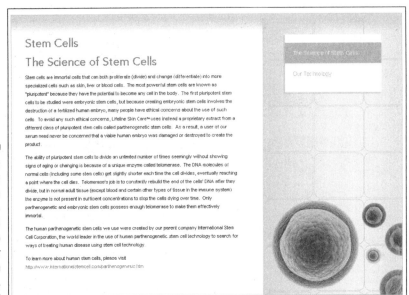

Figure 8-21:
Lifeline
delivers
nanoen-
capsulated
proteins to
keep your
skin
youthful.

Making sunscreens without that icky white stuff

Once upon a time, lifeguards and others who spent a lot of time in the sun would slather on a thick coat of white cream containing zinc oxide, which blocks UV rays but doesn't look that great on your face. Although sunscreens have improved since those days, they can still leave a whitish residue on your skin.

A company called Antaria (www.antaria.com), whose web site is shown in Figure 8-22, uses nanoparticles of zinc oxide to make a sunscreen called ZinClear-IM. This sunscreen protects you from the UV without leaving behind a white coating.

Titanium dioxide is another material used in many sunscreens that can also leave a white residue. BASF is producing powders containing titanium dioxide nanoparticles called T-Lite, for use by sunscreen and cosmetics manufacturers. The titanium dioxide nanoparticles provide protection from UV rays without leaving a white residue.

Chapter 9

Changing the Way We Do Medicine

*O*f all the things nanotechnology makes possible, the promise of making us live longer and healthier lives is probably the most intriguing to many of us. Nanotechnology may be able to extend our lives in two ways. One is by helping to eradicate life-threatening diseases such as cancer, and the other is by repairing damage to our bodies at the cellular level — a nano version of the Fountain of Youth.

Over the last 100 years, our average life span has been increased, in part, by reducing the effect of life-threatening diseases. For example, vaccines have virtually eliminated smallpox. The application of nanotechnology in health-care is likely to reduce the number of deaths over the next decade or so from traditionally difficult-to-eradicate diseases such as cancer and heart disease.

Perhaps the most exciting possibility nanotechnology offers is the potential for dealing with our bodies at the cellular level. As we age, DNA in our cells is damaged by radiation and chemicals in our bodies and our environment. Techniques for building nanorobots are being developed that should make the repair of our cells possible and repair damaged DNA.

In this chapter, we discuss a wide variety of techniques involving *nanomedicine* (the use of nanotechnology for diagnosing, treating, and preventing disease) that promise to make us healthier.

No Lab Test Required: Diagnosing Diseases in the Doctor's Office

Currently, diagnosing a patient's symptoms and pinpointing the disease causing them requires sending a blood or tissue sample to a lab to be analyzed. Typically, the results are returned in about a week, and off you go for a second visit to the doctor.

Nanotechnology can make the process of diagnosing diseases faster and more efficient by enhancing the capability of sensors to identify specific diseases. Soon you won't have to wait a week; the doctor will be able to put a small blood sample in a diagnostic unit in his or her office and get the results in a few minutes. This approach could significantly shorten the time needed to make decisions about how to treat a patient and reduce the need for repeat office visits.

Nanowire-based sensors

Researchers have created sensors by bonding antibodies to nanowires. When biological molecules that indicate the presence of a particular disease attach to matching antibodies, they change the resistance of the nanowire. Because a nanowire is so small, with a diameter of about 10 nanometers, even a small change like bonding an additional molecule to the nanowire can make a noticeable change in its electrical properties.

Using several nanowires in an array, as illustrated in Figure 9-1, with each nanowire functionalized with a different antibody, allows each nanowire to be used to detect a different type of disease.

When you apply a voltage to each nanowire, you drive a current through it. When the resistance of the nanowire changes because the wire has bonded with a virus or protein, the electrical current also changes. This change lets the doctor know which virus or protein is present in the blood sample.

This method is useful when trying to figure out what might be causing certain symptoms in a patient. But a more sensitive test is needed if a disease is in a very early stage, or if a doctor needs to detect any remnants of a particular disease in the patient after therapy is complete. More sensitive diagnostic sensors also use nanowires and antibodies, but these sensors look only for cells carrying that particular disease. To increase sensitivity, you simply use a larger number of those antibodies to increase the chance of detection.

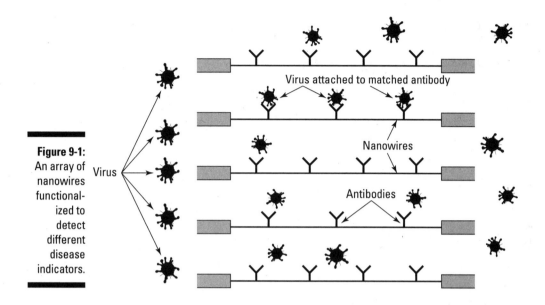

Figure 9-1:
An array of nanowires functionalized to detect different disease indicators.

Virus

Virus attached to matched antibody

Nanowires

Antibodies

Researchers at MIT and Harvard have developed a sensor using billions of carbon nanotubes that they functionalized with antibodies to attach to cancer cells. They've designed this device to be sensitive enough to indicate if even a single cancer cell is trapped on a carbon nanotube. This test would alert doctors that cancer cells are in the bloodstream before the cells can form new tumors. This method could be used, for example, on a patient with a cancer tumor to determine whether the cancer is isolated to that one tumor or is spreading throughout the bloodstream, or to detect HIV being carried in the blood.

Another way to make use of these nanomaterial-based sensors is to determine the level of a particular substance, such as glucose molecules, in a patient. Various researchers are working with either nanowires or nanotubes to make implantable glucose sensors. After the development, testing, and regulatory approvals are complete, patients may be able to wear an unobtrusive patch containing a small insulin reservoir and pump controlled by these nanosensors. This system could automatically inject a diabetic patient with insulin to control the person's blood sugar level.

This method might be useful also for patients with life-threatening allergies to detect an allergic reaction and administer the necessary medicine to halt its progress.

Calling on functionalized quantum dots to find diseases

Quantum dots are semiconductor nanoparticles that glow a particular color when you expose them to light. The resulting color depends upon the size of the nanoparticle. When quantum dots are illuminated by UV light, some of the electrons receive enough energy to break free from the atoms. This allows them to move around the nanoparticle in a *conductance band* in which electrons are free to move through a material and conduct electricity. When these electrons drop back into the outer orbit around an atom (called the *valance band*), they emit light. The color of that light depends on the energy difference between the conductance band and the valance band.

The gap between the valance band and the conductance band, which is present for all semiconductor materials, causes quantum dots to fluoresce. The smaller the nanoparticle, the higher the energy difference between the valance band and conductance band, which results in a deep blue color. For a larger nanoparticle, the energy difference is lower, which results in a reddish glow.

Many semiconductor substances can be used as quantum dots, such as silicon, cadmium selenide, cadmium sulfide, or indium arsenide. Nanoparticles of these or any other semiconductor substance have the properties of a quantum dot.

You can improve the fluorescence of quantum dots by coating them with another semiconductor material. This coating prevents the surface of the quantum dots from being oxidized, which degrades their capability to fluoresce. For example, researchers have found that if they treat a quantum dot made of cadmium selenide (CdSe) with a coating containing zinc sulfide (ZnS), the glow of the quantum dots increases. However, because you also want the quantum dot to mix well with water (because blood is mostly water), a coating of a hydrophilic polymer is added on top of the ZnS. The antibodies that researchers use to attach the quantum dot to diseased cells, such as those in a cancer tumor, are attached to this polymer layer. The resulting structure is shown in Figure 9-2.

Figure 9-2:
Structure of a cadmium selenide–based quantum dot.

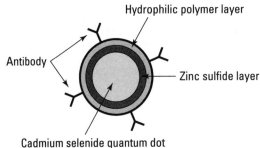

Hydrophilic polymer layer

Antibody

Zinc sulfide layer

Cadmium selenide quantum dot

The fact that quantum dots glow different colors depending on their size is convenient for diagnosing a blood sample to determine if disease indicators are present. Each size of quantum dot is attached to different antibodies. When you place these quantum dots in a blood sample containing a molecule, such as a protein or virus that indicates a particular disease, the quantum dots with the corresponding antibody attach to the protein or virus and form clusters. When you illuminate the solution with ultraviolet light, those clusters glow with the color of that size quantum dot, revealing which virus or other disease indicator is contained in the sample.

Using functionalized iron oxide nanoparticles to spot the culprit

Another way to identify disease indicators in a blood sample is to tag them with magnetic or paramagnetic nanoparticles. The first step in this process is to simply attach antibodies to the magnetic material, for example, iron oxide nanoparticles.

Like iron, iron oxide has magnetic properties. Iron has four unpaired electrons, but in iron oxide, two of those electrons are paired with oxygen, leaving two unpaired electrons. Iron oxide is therefore less magnetic than iron. Because of this lesser magnetic quality, iron oxide is referred to as a *para-magnetic* material.

The paramagnetic properties of iron oxide nanoparticles are not changed from the bulk material except that these tiny particles can go where larger particles never could. These functionalized nanoparticles can attach themselves to a protein or virus, forming a cluster that indicates a particular disease. These clusters of iron oxide nanoparticles can be detected by magnetic resonance imaging (MRI) technology, the same technique used to image organs in your body.

Another approach that is currently being developed to use magnetic nanoparticles to analyze a blood sample is to attach several different antibodies to magnetic sensor tips built on a computer chip. When the chip is placed in a blood sample, any viruses or proteins that match one of the antibodies on the chip would attach to the corresponding antibody. Next, a solution containing magnetic nanoparticles that have been functionalized with antibodies is introduced into the chamber. Those magnetic nanoparticles attach to the virus or protein that attached itself to the antibody on the sensor tip. A signal associated with a particular disease is then sent for that particular virus or protein. This process is illustrated in Figure 9-3.

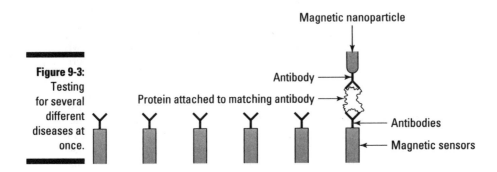

Figure 9-3:
Testing
for several
different
diseases at
once.

Magnetic nanoparticle

Antibody

Protein attached to matching antibody

Antibodies

Magnetic sensors

Enhancing Imaging

Medical imaging has come a long way in the last hundred years. For example, x-rays and MRIs are now commonly used to help doctors pinpoint internal health problems. But the next great improvements in imaging may be largely due to nanotechnology, which helps diagnosticians pinpoint problem spots and improve the quality of images.

Making MRI images crisper with iron oxide nanoparticles

Physicians often use magnetic resonance imaging (MRI) to obtain images of the organs in a patient and avoid potentially harmful imaging methods such as x-raying.

So how does MRI work? Most of the molecules in your body contain hydrogen. Water molecules have two hydrogen atoms, and the organic molecules that make up the rest of our bodies are called hydrocarbons because they contain hydrogen and carbon. The magnetic fields generated by the MRI machine interact with hydrogen atoms throughout the body, producing an image of all the organs.

Step into the Wayback Machine and you may recall those golden moments spent in your high school science class. If you were paying attention, you learned that hydrogen has just one proton in its nucleus. It's this proton in hydrogen that the MRI uses to produce images of the inside of a patient. In the magnetic field generated by the MRI machine, the spin of the protons in the hydrogen atoms are set in one direction.

If you've been unfortunate enough to plow through about 200 advanced mathematics classes to study quantum mechanics, you know that protons have spin. The direction of that spin determines the direction of a magnet, which is composed of the spins of all the charged particles (protons and electrons) together.

To take an MRI image, the MRI machine generates a radio frequency pulse that has just the right amount of energy to flip the spin direction of the protons. When the protons flip back to the spin direction aligned with the magnetic field, they send out another radio frequency pulse. This pulse is detected by the machine, which then uses the pulse to generate an image. The time it takes for the protons to flip back and generate the return radio frequency pulse depends on the protons' location in the body and the density of the tissue at that location. This so-called relaxation time is different for protons in an organ, such as a kidney, than for protons in the bloodstream and is different for healthy tissue than it is for cancer tumors. These differences in the relaxation time are used to generate the MRI images.

By now, you're asking yourself, where do nanoparticles enter the picture? Remember that iron oxide is paramagnetic. You get a better MRI image if paramagnetic nanoparticles are attached to the object you're imaging. Paramagnetic nanoparticles reduce the time it takes for the protons to flip back to the spin direction aligned with the magnetic field. Therefore, the difference in the relaxation time of the tissue that has nanoparticles attached versus the relaxation time of the surrounding tissue is greater, which creates more contrast and produces a clearer image. Because of this effect, researchers are functionalizing iron oxide nanoparticles by coating them with molecules attracted to specific sites, such as cancer tumors, to provide a better MRI image.

Providing fluorescence with silicon quantum dots

Previously in this chapter, we described how quantum dots glow a particular color after you shine light on them and some of the electrons receive enough energy to break free from the atoms. Researchers are developing silicon nanoparticles to be used for fluorescent imaging of diseased tissue in the body, such as tumors.

Because silicon is a semiconductor, silicon nanoparticles are part of the class of nanoparticles called quantum dots. Quantum dots containing some substances, such as cadmium, may be toxic. Researchers believe that quantum dots made of silicon have less chance of producing toxic effects than other types of quantum dots.

Just as they're using cadmium selenide quantum dots for the diagnostic testing of blood samples, researchers are surrounding silicon quantum dots with other layers for the same purpose, as illustrated in Figure 9-4. The first layer is made up of silicon dioxide, which is used to prevent the oxidation of the outer layer of silicon atoms. The second layer is a hydrophilic layer, which allows nanoparticles to mix with water solutions, such as blood. Attached to this polymer layer are the antibodies that attach the quantum dot to diseased cells. Another type of molecule, polyethylene glycol (PEG), is also attached to the polymer layer. As this nanoparticle travels through the bloodstream, white blood cells in the body's immune system might attack the nanoparticle. However, polyethylene glycol molecules shield the nanoparticle from the immune system.

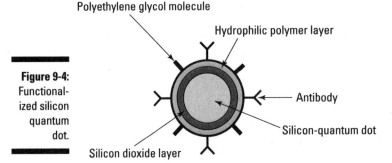

Figure 9-4:
Functionalized silicon quantum dot.

Researchers believe that using quantum dot–based nanoparticles in the body can help them identify smaller cancer tumors and provide better images of cancer tumors. In addition, during surgery, the glowing nanoparticles can help a surgeon identify and remove an entire tumor.

Delivering Drugs More Efficiently

As a society we depend on drugs: drugs to relieve pain, mitigate hay fever, or relax our muscles, for example. What many of us don't realize is that the delivery of those drugs through an injection or by swallowing a pill can be inefficient and even do us damage. Nanotechnology is holding out the promise of some great improvements in the area of drug delivery, from lotions absorbed through the skin to time-released subcutaneous medications.

Breaking through cell membranes

One obstacle to delivering drugs into cells is that many types of nanoparticles can't get through the membranes surrounding cells to deliver drugs. To understand why that's so and how nano can provide a solution for breaking through cell membranes, you have to understand a bit about the nature of cells.

Cell membranes in our bodies are composed of molecules called *phospholipids*. One end of these phospholipids, called the head, is hydrophilic, meaning that it mixes well with water. On the other end of the cell are two tails that are hydrophobic, meaning that they don't mix well with water. These molecules are illustrated in Figure 9-5.

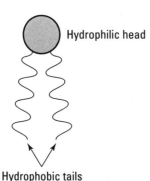

Figure 9-5: The molecules that make up cell membranes.

Hydrophilic head

Hydrophobic tails

The hydrophilic end of the phospholipid mixes well with water because it's polar, meaning that it's composed of atoms that have different levels of attraction to electrons (this measurement of attraction is called *electronegativity*).

The hydrophobic end of the phospholipid is composed of atoms that have similar electronegativity; therefore the electrons are evenly distributed, making this end of the molecule nonpolar. Nonpolar molecules don't mix well with water, a characteristic you've seen in practice if you've ever tried to mix water and oil.

The membrane of a cell is composed of many of these molecules in a two-layer film. In this film, shown in Figure 9-6, hydrophilic ends of the outer layer of the molecules form the outside of the membrane, and the hydrophobic tails of the molecules meet in the middle. This structure has a couple of purposes: the hydrophilic outer layer lets the cell mix with the water-containing fluids in our bodies while the hydrophobic layer in the middle of the membrane prevents the water-containing fluids inside the cell from leaving.

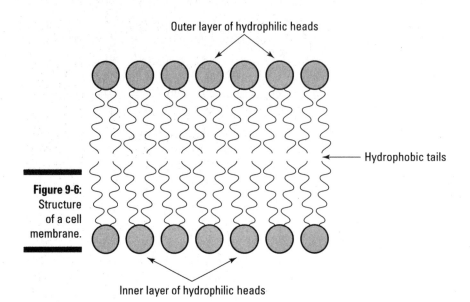

Outer layer of hydrophilic heads

Hydrophobic tails

Figure 9-6:
Structure
of a cell
membrane.

Inner layer of hydrophilic heads

The result of this structure is that the cell membrane blocks the entry of many therapeutic drugs into the interior of the cell, motivating researchers to develop methods to deliver these drug molecules through the cell membrane.

One way to get drugs through cell membranes is to enclose drug molecules in artificially created spherical nanoparticles called *liposomes.* You place these molecules with the hydrophilic head and hydrophobic tails in water that contains the drug molecules you want to encapsulate. The hydrophilic heads line up to form an outer shell facing the water solution, while another set of hydrophilic heads lines up to form an inner shell that contains a solution of drug molecules, as illustrated in Figure 9-7.

Figure 9-7:
Spherical
liposome
nanoparticle
enclosing
drug
molecules.

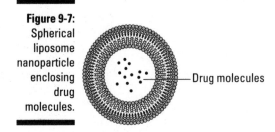

Drug molecules

The shells of these liposomes can fuse with cell membranes. If a therapeutic drug is encased in a liposome, the liposome membrane creates an opening as it fuses with the cell membrane, and the drug inside the liposome can be delivered into the cell, as illustrated in Figure 9-8.

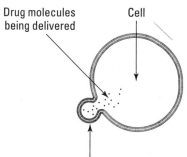

Figure 9-8: Liposome nanoparticle delivering drug molecules in a cell.

Drug molecules being delivered

Cell

Liposome fused with cell membrane

Proteins are designed to deliver materials through cell membranes. For example, hemoglobin is a protein that picks up oxygen in your lungs and distributes it to cells throughout your body. Hemoglobin is actually capable of burrowing through cell membranes to deliver oxygen. In a different approach to drug delivery, AbraxisBioScience is enclosing therapeutic drug molecules in an albumin protein. Nanoparticles of albumin can burrow through cancer cell membranes and deliver a drug to the interior of the cell.

Another method is a little more aggressive, kind of like the Bruce Lee of drug delivery. This method, developed at the Georgia Institute of Technology, blasts temporary holes in the cell membrane, allowing therapeutic drug molecules to enter the cell. In this method, researchers inject carbon nanoparticles into the fluid floating around cancer cells, and then a laser heats the fluid. This heat creates gas bubbles; when the bubbles burst, they blow a hole in the membrane of the cancer cells. Drug molecules floating in the fluid around the cancer cells can then enter the cells and destroy them.

Targeting the right spot

As we discussed in the previous section, nano can help to get therapeutic drugs through cell walls, but how can it help researchers find the right cells to target? Just as diagnostic sensors use certain proteins that indicate the presence of a particular disease and attach antibodies to them (see the earlier section, "Nanowire-based sensors"), you can improve the delivery of drugs by matching antibodies to certain proteins. This method, called *targeted drug delivery,* involves the delivery of drugs directly to diseased cells within the body. Researchers are trying to identify which antibodies can be attached to drug-carrying nanoparticles to functionalize them so that they attach to a protein or another biological molecule on a diseased cell.

Nanoparticles have been functionalized in this way by various researchers and companies to target different types of cancers. By using therapeutic drug-carrying nanoparticles, chemotherapy drugs can be delivered directly into cancer cells. This method should stop or significantly reduce the damage that less targeted chemotherapy drugs cause to healthy cells.

Another use of functionalized nanoparticles for targeted drug delivery may help researchers tackle arteriosclerosis caused by plaque building up inside arteries. The portion of plaque most likely to flake off and cause a blockage downstream is on the edge of the plaque buildup, where the plaque meets the artery wall. Researchers have determined how to target nanoparticles to these critical regions of plaque buildup so that therapeutic drug molecules are delivered where they can be most effective.

One problem with having nanoparticles searching out diseased cells floating through the bloodstream of a patient is that white blood cells in the immune system attack foreign materials in the bloodstream. To solve this, researchers attach molecules called polyethylene glycol (PEG), which hide the nanoparticles from the immune system. The PEG allows the nanoparticles to flow through the bloodstream without being detected.

Multitasking nanoparticles

A next logical step in developing drug-delivering nanoparticles is to find a way for them to perform multiple functions. For example, researchers at UCLA and Yonsei University in Korea have developed nanoparticles containing an iron oxide core, a nanoporous silica region, chemotherapy drugs stored in the silica nanopores, and targeting antibodies to bond the nanoparticles to cancer cells.

The iron oxide core has two duties. First, it allows an MRI to provide an image showing that the nanoparticles are attached to the targeted cancer cells. Then, when researchers apply an oscillating magnetic field, the iron oxide core heats up and causes the silica nanopores to release the chemotherapy drug into the cancer cells.

Making the daily dose obsolete

You know those little pill holders with compartments for vitamins, hay fever pills, blood pressure medication, and so on that you see more mature citizens carting around with them? In our healthcare conscious world, it's hard to avoid being on some medication or preventive nutritional supplement at some point in your life. Wouldn't it be lovely if you could take a pill once every year instead of every single day? That scenario just might be possible. Various researchers are developing nanoporous drug delivery particles.

These nanopores are filled with therapeutic drugs that can be released over a period of months or years.

For example, a company called pSivida Limited offers a drug delivery product called BioSilicon, a silicon particle riddled with nano-sized pores. The drug is loaded into the pores. As the silicon particle dissolves, the particle releases the drug. pSivida can customize the size and porosity of silicon particles to control the time it takes them to dissolve. BioSilicon may be used in implants under the skin that could release a drug over days, weeks, or months.

Researchers are also developing implantable drug delivery products using other types of nanomaterials, such as nanoporous silicon dioxide or titanium dioxide nanotubes. Researchers are investigating various applications for these materials. For example, one study shows that a time-release material injected into the eye can supply therapeutic drugs to help combat blindness in diabetic patients. Another study on lab animals has shown that time-release therapeutic material injected into the brain can help to reduce the severity of seizures in epileptics.

Stopping flu in its tracks

What can nanotechnology do about the Holy Grail of medical breakthroughs, preventing the common cold, or it's equally unpleasant cousin, the flu? The immune system in the lungs produces a material called inducible bronchus-associated lymphoid tissue (iBALT) in response to a flu virus. Although the role of iBALT in fighting diseases is not completely understood, researchers at Montana State University are working on protein nanoparticles that, when inhaled by lab animals, activate the same immune response system that nature provides in your lungs to provide protection from viruses such as influenza for about a month.

This method offers us the prospect of minimizing the effect of the flu or colds regardless of the particular virus causing the disease, unlike a vaccine that is effective only against the particular virus it's designed to fight. This particular breakthrough is still at the stage of testing on lab animals (which are sneezing and coughing as we speak just to make our futures brighter).

Going skin deep

One of our favorite medical advances portrayed on the original *Star Trek* series was when Dr. McCoy pressed a device against a patient's arm to deliver a drug dose with a little "whoosh" and no accompanying "ouch." Drug delivery that avoids painful injections is becoming more of a reality thanks to nano.

Researchers at the University of Michigan are working on the *nanoparticle field extraction thruster* (nanoFET), developed to provide propulsion for spacecraft. This device uses nanoparticles that get charged when they lose electrons as they come into contact with an electrode at a positive voltage. After the nanoparticles are charged, they can be accelerated by an electric field, providing thrust to a spacecraft.

Researchers think there could be a method to deliver nanoparticles that contain therapeutic drugs using the nanoFET. Researchers believe that nanoparticles coming out of the nanoFET will just pass through a patient's skin into the bloodstream. This application is just an idea at this time, but it's one whose outcome we eagerly anticipate.

Another advance in painless drug delivery is happening as various researchers are encapsulating drugs in emulsion nanoparticles that transport the drug through your skin; you simply spread the emulsion, as you do hand lotion, on your skin. The nanoparticles are small enough that they are simply absorbed through the skin. The emulsion provides a reservoir of the drug just under the surface of your skin from which the drug can continue to move into your bloodstream, maintaining stable levels of the drug over time.

This method avoids passing the drug through your stomach with the associated side effects (for example, some pain relief drugs can cause ulcers when taken orally) and also frees you from the pain of getting an injection or the discomfort of a patch on your skin.

This technique may eventually be used to deliver a range of drugs, such as hormones, pain killers, allergy medications, and arthritis and cancer treatments.

Treatments

When doctors identify diseased cells such as those in a cancer tumor, they have certain treatments at their disposal. Unfortunately, some of those treatments do much harm to patients in the attempt to kill off the disease. Nanotechnology offers some clues for how to target treatments and avoid damaging healthy tissues and systems, and even emulate some of the ways in which our own immune system keeps us healthy.

Zapping diseased cells with heat

Sadly, many of us have seen the debilitating effects of currently available radiation therapies on the patients undergoing them. The problem with radiation therapy is that the radiation can cause severe side effects by damaging surrounding healthy tissue while trying to target diseased tissue.

Researchers are therefore looking for ways to replace radiation therapy. An interesting alternative is hyperthermia therapy.

Hyperthermia therapy raises the temperature of diseased tissue, such as cancer tumors, to kill it off. Researchers have found that raising the temperature of cells above 45 degrees centigrade does the trick. Nanoparticles are used to absorb energy from sources such as infrared light and convert that energy into heat, which is then applied directly to diseased cells with no or little damage to surrounding tissue.

Several researchers are using gold nanoparticles for hyperthermia therapy because gold has the capability to convert certain wavelengths of light into heat. As with all metals, gold contains electrons that are not tied to any particular atom but are free to move throughout the material. These electrons help to conduct a current when you apply a voltage. Depending on the size and shape of the nanoparticles, these free electrons absorb the energy from a particular wavelength of light. At the right wavelength, light makes the cloud of free electrons on the surface of the gold nanoparticles resonate, heating them and transferring that heat to the target cells.

Two types of gold nanoparticle shapes are most efficient in converting light into heat:

- ✔ **Gold nanorods:** These solid cylinders of gold have a diameter as small as 10 nm. By using nanorods with various combinations of diameter and length, researchers can change the wavelength of light that the nanorod absorbs.

- ✔ **Nanoshells:** These types of nanoparticles consist of a gold coating over a silica (silicon dioxide, the same material as glass) core. By using nanospheres with variations in the thickness of the gold coating and the diameter of the silica core, researchers can change the wavelength of the light that the nanosphere absorbs.

Various researchers are using nanorods, nanoshells, or other nanoparticles that convert light to heat (such as carbon nanotubes) to develop methods for localized heat treatment of diseased regions of the body.

For more about the nature and use of gold nanoparticles, see Chapter 3.

An interesting alternative to hyperthermia therapy uses nanoparticles composed of titanium dioxide functionalized by attaching an antibody that is attracted to cancer cells. When researchers shine visible light on the cancer tumor, the titanium dioxide, which is a photocatalyst, donates electrons to oxygen in the bloodstream, creating negatively charged oxygen atoms. These oxygen atoms react with molecules in the cancer cells, killing them. Because visible light can't penetrate very far into the body, this method will work only for cancer cells close to the surface.

Combating infection with antimicrobial treatments

From falling off our tricycles when we were three to cutting our fingers with a kitchen knife, we've all experienced wounds — and they're no fun. One of the biggest dangers with wounds, however, is not the initial pain but the possibility of infection. Nanotechnology is being used to fight harmful bacteria that causes infection in wounds.

One of the earliest nanomedicine applications was the use of nanocrystalline silver as an antimicrobial agent for the treatment of wounds.

For example, researchers have demonstrated that a nanoparticle cream can help to fight staph infections. The nanoparticles contain nitric oxide gas, which is known for its capability to kill bacteria. Studies on mice have shown that using the nanoparticle cream to release nitric oxide gas at the site of staph abscesses significantly reduces infection.

Another useful application of this technology is burn dressings coated with nanocapsules containing antibiotics. If an infection begins, the harmful bacteria in the wound causes the nanocapsules to break open, releasing the antibiotics. This method allows much quicker treatment of an infection and reduces the number of times you have to change a dressing.

The elimination of bacterial infections in a patient within minutes by using nanorobots instead of delivering treatment with antibiotics over a period of weeks is very promising. For more about the future of this treatment, check out the section on nanorobots later in this chapter.

Taking action when our immune systems turn against us

Type 1 diabetes, an autoimmune disease, is a serious health problem. An autoimmune disease involves the immune system mistakenly attacking functional cells in the body. Type 1 diabetes is caused by T cells, a certain type of white blood cell that is part of the body's immune system. In those with type 1 diabetes, these white blood cells destroy the beta cells in the pancreas so that it cannot produce insulin.

A vaccine generally causes the immune system to produce white blood cells to fight off disease. A researcher at the University of Calgary has developed a nanoparticle that prevents these white blood cells from attacking the beta cells. Attached to the surface of these nanoparticles are segments of beta cells that alert the immune system to produce a different type of white blood cell that stops the T cells from attacking the beta cells, acting like a vaccine.

This method has so far been tested only on lab animals, where it has prevented the onset of diabetes in prediabetic mice and returned blood sugar levels to normal in diabetic mice.

Sequencing DNA

You know that your DNA predetermines some possible health risk factors. Determining the structure of your DNA can help determine what diseases you might be susceptible to and helps your doctor and you to take preventative measures.

Determining a person's DNA involves finding the sequence of the molecules, called bases, which connect single strands of DNA into a double helix. Four kinds of molecules are involved: adenine, thymine, guanine, and cytosine. The order of these molecules determines the genetic structure of an individual. The process of figuring out the order of these molecules is called *DNA sequencing.* Researchers are developing faster and less expensive methods for performing DNA sequencing.

When DNA strands are fed through a nanopore with a voltage difference across the pore, you can identify each molecule in the DNA strand by the amount of current that flows across the nanopore.

The main problem with this method is that the distance between bases in the DNA strands are about half a nanometer. Therefore, if the nanopore is thicker than half a nanometer, you can't measure individual bases in the DNA strand. For that reason, researchers must use material that is only one atom thick to make up the nanopore. Graphene sheets come to the rescue. These sheets are only one carbon atom thick, so nanopores made of graphene are thin enough to resolve individual bases in DNA strands.

Using nanorobots

Like us, you may have grown up with robots in science fiction movies and books. If you think robots are cool, you'll be interested to hear that researchers are developing robots about the size of the cells in our bodies that have a propulsion system, sensors, manipulators, and even an onboard computer that can perform tasks on nanoscale objects. Nanorobots are definitely not ready for prime time, but Figure 9-9 shows an artist's conception of nanorobots at work zapping pathogens in the bloodstream.

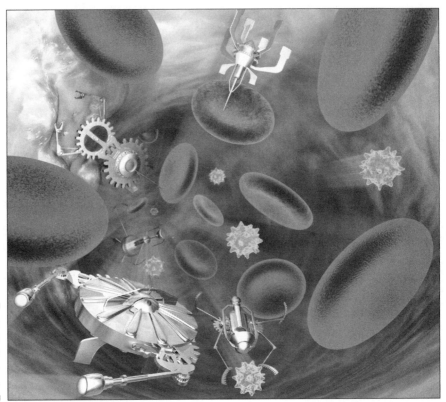

Figure 9-9:
Nanorobots
at work in
the blood-
stream.

The various methods being developed for building nanorobots are discussed in Chapter 4.

Here are a few of the nanorobots that are being developed for use in medicine:

✔ **Microbivore nanorobots:** These nanorobots would function similarly to the white blood cells in our bodies, but they are designed to be much faster at destroying bacteria. This type of nanorobots should be able to eliminate bacterial infections in a patient within minutes, as opposed to the weeks required for antibiotics to take effect. Microbivore nanorobots are designed so that antibodies attach to the particular bacteria the robot is seeking. After bacteria attaches to an antibody, an arm grabs the bacteria and moves it to the inside of the nanorobot, where it's destroyed. Bacteria is then discharged into the bloodstream as harmless fragments.

One of the pioneers of nanomedicine is Robert Freitas, who has published design studies for the nanorobots discussed here. You can see his design study for the microbivore nanorobot at www.rfreitas.com/Nano/Microbivores.htm.

✔ **Respirocyte nanorobots:** These nanorobots would function in a similar way to the red blood cells in our bodies; however, they are designed to carry much more oxygen than natural red blood cells. This design could be very useful for patients suffering from anemia. These respirocyte nanorobots would contain a tank in which oxygen is held at a high pressure, sensors to determine the concentration of oxygen in the bloodstream, and a valve that releases oxygen when sensors determine that additional oxygen is needed.

✔ **Clottocyte nanorobots:** These robots function similarly to the platelets in our blood. Platelets stick together in a wound to form a clot, stopping blood flow. Depending on the size of the wound, significant blood loss can occur before a clot is formed. A system of clottocyte nanorobots would store fibers until they encounter a wound. At that point, the nanorobots would disperse their fibers, which would then come together to create a clot in a fraction of the time that platelets do.

✔ **Cellular repair nanorobots:** These little guys could be built to perform surgical procedures more precisely. By working at the cellular level, such nanorobots could prevent much of the damage caused by the comparatively clumsy scalpel.

Cellular repair

Perhaps the most exciting of these possibilities is the potential for repairing our bodies at the cellular level. Techniques for building nanorobots are being developed that should make the repair of our cells possible. Nanorobots would be able to repair damaged DNA and allow our cells to function correctly.

This capability to repair DNA and other defective components in our cells goes beyond keeping us healthy: It has the potential to restore our bodies to a more youthful condition. This concept is discussed in Eric Drexler's *Engines of Creation*. Drexler states:

> *"Aging is fundamentally no different from any other physical disorder; it is no magical effect of calendar dates on a mysterious life-force. Brittle bones, wrinkled skin, low enzyme activities, slow wound healing, poor memory, and the rest all result from damaged molecular machinery, chemical imbalances, and mis-arranged structures. By restoring all the cells and tissues of the body to a youthful structure, repair machines will restore youthful health."*

Replacing problematic DNA

Various types of proteins are capable of repairing cells. Proteins actually cut out segments from defective DNA strands, and then other proteins regrow the DNA strand to match neighboring, nondefective DNA.

The Nanomedicine Center for Nucleoprotein Machines is one organization dedicated to finding ways to treat damaged DNA. Instead of building cellular repair robots from scratch, this center is working on reengineering the proteins that currently repair defective DNA to produce nanomachines that can do the same work. In fact, the *Nucleoprotein Machines* in the Center's name relates to the fact that proteins are Nature's workhorses for repairing damaged DNA.

Nucleoprotein is the term for a protein that is attached to a nucleic acid, such as DNA.

The Center is testing its methods on sickle-cell disease in mice. Sickle-cell disease is a genetic disease that causes red blood cells to have an abnormal shape that moves through the bloodstream less easily than normally shaped red blood cells. This condition causes anemia. Sickle cell is a painful disease that shortens life and currently has no known cure. The Center uses mice and models of sickle-cell disease to find a way to repair the mutated gene that causes it.

Regenerating cells

Stem cells can produce new cells that in turn regenerate many different types of damaged cells. Various researchers are working on methods for regenerating damaged organs or nerves by placing stem cells in damaged tissue.

Simply injecting stem cells into damaged tissue is inefficient because stem cells can wander from the spot where they're placed. Researchers are finding that therapy is much more effective if they attach the stem cells to nanofibers and then inject the nanofibers at the precise location of the damaged cells.

This method could be very helpful to patients recovering from heart attacks. Heart attacks damage heart cells, reducing the capability of the heart to pump blood to the rest of the body. Normally, these damaged cells cannot recover, which leaves the option of heart transplants when enough damage has occurred. Experiments with lab animals have shown that damaged hearts show significantly better recovery after being treated with nanofibers to which stem cells have been attached. The nanofibers are made of biodegradable material that dissolves, leaving behind regenerated tissue.

Researchers call these nanomaterials *scaffolds*. Different types of nanomaterial are being used as scaffolds. Some researchers are using nanofibers made of polymers, and other research groups are using nanofibers made of biological molecules, such as peptides or proteins.

These scaffolds could be applied in various ways. For example, researchers working on regenerating heart tissue are injecting nanofibers and stem cells that self-assemble into a scaffold. Another research group working on regenerating retinal cells is using a surgical procedure to implant a disk-like scaffold into the eye. To regenerate retinal cells to try to cure macular degeneration, researchers have found that they need to use this scaffolding to guide stem cells to let them know where to grow the new retinal cells. At this time, researchers are working with scaffolds at the micro level. The plan is to make scaffolds with details on the nano level, which will make it possible to give new cells more guidance in forming their connections with existing cells in the eye.

Although this method is now being tested in lab animals and is years from being available for use in human patients, it has the potential for regenerating many types of tissue. Applications of this method could regenerate damaged spinal cords, cartilage in joints for relief of arthritis, and nerve cells in the brain to treat Parkinson's disease.

For the latest applications of nanotechnology in medicine, visit Understanding Nano's nanomedicine page at `www.understandingnano.com/medicine.html`.

Chapter 10

Saving Energy with Nano

*O*ne of the fathers of nanotechnology, Dr. Richard Smalley, believed that nanotechnology might help provide a solution to the growing energy demands of the world. According to Smalley:

> *"Energy is the single most important problem facing humanity today — not just in the U.S. but also worldwide. The magnitude of this problem is incredible. Energy is the largest enterprise on Earth — by a large margin. . . . While conservation efforts will help the worldwide energy situation, the problem by mid-century will be inadequate supply."*

Nanotechnology can help address this huge challenge by improving the way energy is generated, distributed, and stored. In this chapter, we cover what's going on in nanoenergy, as well as review some future directions in nanoenergy being explored by researchers, governments, and businesses.

Exploring the Big Three of Energy: Generation, Distribution, and Storage

Smalley's vision of how nanotechnology could improve our energy challenges is playing out before our eyes. Today, nanotechnologists are exploring the role of nanotechnology in making energy more efficient in three key areas:

✔ **Energy generation:** Improvements in the generation of energy consist of solar cells that cost less and are more efficient and the production of new materials for more efficient fuel cells. Because solar cells and fuel cells have traditionally been costly to produce and use, these efficiencies could make them both finally practical.

✔ **Electricity distribution:** Very low resistance electric transmission wires could improve electricity distribution. These wires would contain carbon nanotubes, which have significantly less resistance than conventional wires. This more efficient distribution grid would allow electricity generated at power plants to be transported thousands of miles with very little power loss. As a result, power could be generated at the most logical sites; for example, solar power in deserts, geothermal power at geyser locations, and wind power in mountain passes.

✔ **Energy storage:** The use of more efficient batteries has improved the storage of energy. Improved batteries or other devices, such as ultracapacitors, could be used to store energy locally for as long as 24 hours. The capability to store energy locally would allow communities or individuals to buy power at the cheapest time of day and to never again suffer from those annoying short-term power outages.

In the rest of this chapter, we look more closely at how nano is helping to make these improvements possible.

If you're interested in learning more about Smalley and his vision of nanotechnology, check out this video on YouTube: www.youtube.com/ watch?v=eR9uHR7uAX4&NR=1.

Generating Energy More Cheaply

Both solar cells—the kind you see plastered on neighborhood roofs—and fuel cells—like those being developed for use in electric cars—hold promise for improving our energy generation equation. However, both aren't ready for prime time because they aren't efficient enough or cost-effective enough to outdo traditional sources of energy. Nano may help them turn the corner on cost and efficiency.

Souping up solar cells

Solar cells are one of the great hopes of those seeking more environmentally friendly power, so they're a logical first topic for our discussion of nano and energy. These intriguing devices allow us to tap into the powerful energy generated by our sun. But how do these devices work and how could nano make them work even more efficiently?

Sunlight hitting the Earth has about 10,000 times the energy used by the Earth's population, according to a calculation by scientists at NASA. Therefore, solar cells have the potential for generating much of the electricity we'll need in the future.

The inner workings of a solar cell

Light is produced in a bulb by running an electric current through it. A solar cell is just the opposite: light shining on a solar cell produces an electric current. Figure 10-1 shows the structure of a conventional solar cell.

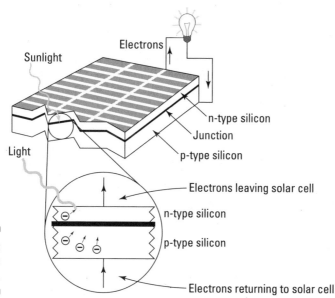

Figure 10-1:
Inside a
solar cell.

A solar cell's key ingredient is a slice of semiconductor, often made up of crystalline silicon. Different impurities are added to the top and bottom layers of the slice (a process called *doping*) to enable the solar cell to produce electricity. The impurities added to the top layer (for example, atoms of arsenic) have five electrons available for bonding to other atoms, instead of the four electrons in the surrounding silicon. The impurities added to the bottom layer (for example, atoms of boron) contain three electrons, one less than the four electrons in the surrounding silicon atoms. The top layer, to which the five-electron impurity has been added, is called *n type* because of the extra electrons, which carry negative charges. The bottom layer, to which the three-electron impurity has been added, is called *p type* because the shortage of electrons creates holes that are *positive* charge carriers.

This tasty sandwich of different types of silicon in a solar cell sends electrons (that is, an electric current) through wires to power various gadgets when light hits the cell. Because each solar cell produces only about half a volt, manufacturers combine solar cells into panels to produce a heftier voltage.

But solar cells are costly to produce and install, and the energy used to create solar cells is currently more than their typical energy savings over several years. Many manufacturers are reducing the cost of solar cells by reducing the thickness of the semiconductors they contain. This type of solar cell is called a *thin-film solar cell*, also referred to as a *second-generation solar cell.*

Bringing big improvements to solar cells with nano

The cost of electricity produced by conventional solar cells is currently higher than the cost of electricity produced by power plants that use coal, gas, oil, or water. To change this equation, we need to reduce the cost of manufacturing (both in terms of dollars and energy), reduce the cost of solar cell installation, and dramatically increase efficiency of solar cells. Nanotechnology may be able to do all that.

Solar cells that use nanotechnology, which are considered third-generation solar cells, should be able to generate electricity at considerably lower cost, but they aren't as efficient as first-generation solar cells. When the efficiency of nano-based solar cells improves and the cells can be manufactured in high volume, they should be more cost-effective than conventional methods of generating electricity, such as coal or gas.

Using nanoparticles in the manufacture of solar cells could help make them a more viable option in several ways. For example, the slices of semiconductor crystal used in conventional solar cells are expensive to generate because they are made with costly vacuum equipment using high temperatures. One way to reduce the production cost of semiconductor material is to use *semiconductor nanoparticles.* The nanoparticles are printed as part of an ink solution on metal foil in a process that results in a semiconductor layer. This process uses much less heat and much less expensive equipment, so you get much cheaper solar cells.

In addition, various companies are working on solar cells that don't use semiconductor layers. Instead the n and p layers are made from different organic molecules that allow the cell to function like silicon-based solar cells. Organic solar cells can be manufactured for significantly less than traditional solar cells. In addition, the finished cells are housed in rolls of flexible plastic instead of rigid crystalline panels, which may reduce the cost of installing them in buildings. Nanomaterials have improved the interface between the n and p layers in these organic solar cells so that electrons can escape the solar cell and move into wires more efficiently. This approach can raise the efficiency of the cells from 1 percent to between 6 and 8 percent.

Nanotechnology applications under development for solar cells

Nanotechnology is a fast-moving field, so what is hot today might be old news tomorrow. That said, several intriguing examples of how nano is improving solar cells may help you understand the types of advances in the field of energy that could be possible.

Companies such as International Solar Electric Technology and Nanosolar are involved in making lower-cost solar cells using semiconductor nanoparticles (see the preceding section). By 2012, Nanosolar intends to be capable of making enough solar panels to generate more than 100 megawatts. The company states that their solar cells will replace the energy used to manufacture them in a few months, rather than the years it takes to recover the energy used in making a conventional solar cell.

Konarka Technologies, Global Photonic Energy Corporation, Solarmer Energy, and others are working on developing organic solar cells, which, as mentioned in the preceding section, have a more efficient interface between the n and p layers. Their products could improve the efficiency of organic solar cells from 1 percent to about 6 to 8 percent. Although this is only about half the efficiency of conventional solar cells, organic solar cells have greater flexibility. Konarka Technologies is currently supplying solar cells using this technology to manufacturers of backpacks that charge cell phones.

Researchers at the New Jersey Institute of Technology have demonstrated that combining carbon nanotubes and buckyballs in an organic solar cell can increase the chance that electrons will escape the solar cell before being absorbed by a molecule in the organic layer. In this process, the buckyballs and nanotubes combine in a snake-like structure. When sunlight hits the organic molecules, it generates an electron. With luck, the electron will encounter a buckyball. The electron is then absorbed by the buckyball and travels down the carbon nanotube to the electrode on the outside of the solar cell and then to the wires connecting the solar cell to the device to be powered.

The nano brain trust at Stanford University has also found a way to trap light in organic solar cells for a longer period of time. By using a nanoscale organic layer, which is much thinner than the wavelength of light, the light stays in the solar cell longer and excites more electrons, producing more electricity.

Firing up fuel cells with nano

You may be looking forward to the day when cars powered by hydrogen fuel cells replace gasoline-powered cars, but it may take a while. The major obstacles to widespread use of cars powered by hydrogen fuel cells are the lack of a network of hydrogen fuel stations, the high cost of hydrogen fuel cells, and the need for lightweight and safe hydrogen fuel tanks.

Nanotechnology can't create a chain of fuel stations, but it may help with the cost of the fuel cells and make it possible to make lighter weight, safer fuel tanks.

Before hydrogen-powered cars become a reality, you may see small fuel cells powering portable devices such as laptops. These small fuel cells are called DMFCs (direct methanol fuel cells) because they usually get around the problems inherent in storing hydrogen by using methanol as a fuel. DMFCs last longer than batteries. Rather than plugging your device into an electrical outlet and waiting for the battery to charge, you simply insert a new cartridge of methanol into the DMFC device and you're ready to go. As with hydrogen fuel cells, nano may be able to reduce the cost of producing DMFCs.

The anatomy of a fuel cell

A fuel cell combines hydrogen and oxygen in a chemical reaction that produces electricity, heat, and water, but no pollution.

Mixing hydrogen and oxygen can be a dangerous business (think of the Hindenburg). Fuel cells mix hydrogen and oxygen in a safe and controlled fashion, and much of the energy produced takes the form of electricity rather than heat.

Figure 10-2 shows the structure of a fuel cell. Essentially, here's what goes on in a fuel cell: A catalyst causes a reaction that breaks down hydrogen and oxygen to ions and electrons. Some of these ions move through a membrane to the other side of the cell while the electrons move through a wire and produce power.

The distribution hurdle

Car companies are producing a small number of fuel cell–powered cars, but widespread use of cars powered by hydrogen fuel cells won't happen until refueling stations become as ubiquitous as your neighborhood gas station.

Honda is leasing the fuel cell–powered FCX Clarity FCEV to about 200 customers in southern California, where they have arranged for some hydrogen refueling stations. The State of California estimates that about 45,000 passenger cars powered by fuel cells will be sold in California in 2017. To give the car owners a place to refuel, California is helping to fund some retail hydrogen fuel stations in major cities. They estimate that they need to add ten new stations a year between now and 2017 to provide enough capacity for the hydrogen-powered cars that will be on the streets at that time.

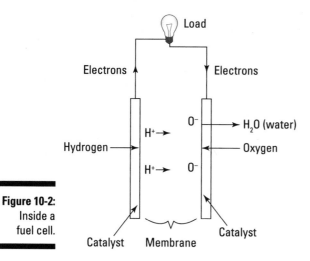

Figure 10-2:
Inside a
fuel cell.

Improving fuel cells with nano

Nano can improve the efficiency of fuel cells in several ways. As mentioned in the preceding section, fuel cells require the movement of ions through membranes. By using nanopores, you can limit what gets through the membranes and better control the reaction that occurs in the fuel cell to produce energy. Researchers are using nanotechnology to change the characteristics of these nanopores and the acid solution inside them. For example, researchers have capped the ends of nanopores to trap the acidic solution inside the membrane, thus improving the transport of hydrogen ions through the membrane in low humidity. This capability opens up the possibility of making fuel cells that operate in a wide range of humidity conditions.

A catalyst, usually platinum, makes fuel cell reactions at lower temperatures occur more easily. A second way that nanotechnology can improve the efficiency of fuel cells involves improving the catalyst. Researchers use nanoparticles to increase the surface area available for reactions, thus making the reaction more efficient — and less costly because less platinum is needed.

Finally, nano may help produce hydrogen storage tanks that are small enough and lightweight enough to use in cars. Hydrogen bonds easily to carbon, so researchers have investigated the storage of hydrogen in *graphene* (carbon sheets). Because graphene is only one atom thick, it has the highest surface area exposure of carbon per weight of any material. High hydrogen-to-carbon bonding energy and carbon's high surface area exposure make graphene a good candidate for storing hydrogen more efficiently.

Nanotechnology applications under development for fuel cells

Several groups are exploring the use of nanotechnology to improve the efficiency of fuel cells. Researchers at the SLAC National Accelerator Laboratory have developed a way to use 80 percent less platinum for the cathode in fuel cells, which could significantly reduce their cost. The researchers alloyed platinum with copper in nanoparticles and then removed the copper from the surface of the nanoparticles, causing the platinum atoms to move closer to each other. The reduced spacing between atoms (called *lattice spacing*) changes the electronic structure of the platinum atoms so that the separated oxygen ions are more easily released, making the catalyst more effective. A more effective catalyst also means that less catalyst is required.

Another way to reduce the use of platinum as a catalyst in fuel-cell cathodes is being developed by researchers at Brown University. They deposited a one-nanometer thick layer of platinum and iron on spherical nanoparticles of palladium. In laboratory tests, they found that a fuel cell using a catalyst made with these nanoparticles generated 12 times more current than one containing a catalyst using pure platinum. The fuel cell also lasted ten times longer. The researchers believe that this improvement is due to a more efficient transfer of electrons.

Hydrogen produced from hydrocarbons, such as oil or natural gas, contains carbon monoxide. To use this hydrogen in fuel cells, you have to remove that carbon monoxide, which increases the cost. Researchers at Cornell University have found a way to reduce the amount of platinum used in the fuel cell and to increase the tolerance of the fuel cell for some contaminates in the hydrogen fuel, which decreases the cost of producing the hydrogen.

The researchers prepared a catalyst composed of platinum nanoparticles on a support with titanium oxide tungsten. They compared the performance of this catalyst in fuel cells to one composed of pure platinum when using hydrogen contaminated with 2 percent carbon monoxide. The performance of fuel cells using the platinum catalyst dropped by 30 percent, but the performance of the fuel cells using the platinum nanoparticle catalyst dropped by only 5 percent.

Researchers at the University of Illinois have developed a proton exchange membrane using a silicon layer with pores that are about 5 nanometers in diameter and capped by a layer of porous silica, as shown in Figure 10-3. The silica layer ensures that water stays in the nanopores. The water combines with the acid molecules along the wall of the nanopores to form an acidic solution, providing an easy pathway for hydrogen ions through the membrane. This membrane had much better conductivity of hydrogen ions (100 times better conductivity was reported) in low-humidity conditions than the membrane normally used in fuel cells. This approach could result in the creation of fuel cells that operate in environments with a wide range of humidity.

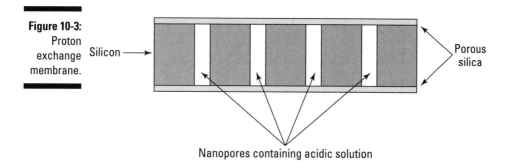

Figure 10-3:
Proton
exchange
membrane.

Silicon →

Porous
silica

Nanopores containing acidic solution

Improving Energy Distribution

Generating energy is just the first step in an energy system. The second important step is getting that energy to those who need it. Nano is helping the electric transmission grid in several ways.

Making wires sing

The electrical resistance in conventional electric transmission lines causes power losses when converting electrical power into heat. Reducing the electrical resistance would cut the losses in transmission lines and allow electricity from power plants to be used in more widespread locations.

Nanotechnology applications under development for distribution

Researchers at Rice University are working to develop wires containing carbon nanotubes that would have significantly lower resistance than the wires currently used in the electric transmission grid. The so-called *armchair quantum wire* would be composed of carbon nanotubes woven into a cable. The term *armchair* comes from the armchair shape, shown in Figure 10-4, seen in the lattice of metallic-type carbon nanotubes. The term *quantum* is used because electrons move from one carbon nanotube in the wire to the next by a quantum mechanical method called *tunneling*. This method is the quantum analogy of walking toward a hill and finding a tunnel that allows you to pass under the hill, saving you the effort of climbing over it.

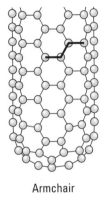

Figure 10-4:
Carbon atoms in metallic-type nanotubes align in an armchair shape.

Armchair

When carbon nanotubes are created, they appear as a tangled web, as shown in Figure 10-5. One challenge in making quantum wire is that the carbon nanotubes must be untangled so they can be woven to form the wire. Recently, researchers at Rice University have made improvements in separating nanotubes using an acid to produce fluid *dopes* (a solution), similar to the dopes used in the industrial spinning of fibers.

Figure 10-5:
Newborn nanotubes are tangled.

Courtesy of Arkema

Storing Energy More Efficiently

After you've generated and distributed energy, you have to store it. Batteries and capacitors play their part as energy storage devices. Nano has important contributions to make to the efficiency of energy storage. To understand how, you need to know how batteries work.

Understanding how batteries work

Your garden-variety battery has multiple cells, with each cell containing an electrolyte, an anode, and a cathode, as shown in Figure 10-6. When you turn on a battery-powered gadget, the voltage between the anode and cathode in the battery causes electrons to flow through the wires to provide power. At the same time, *positive ions* (atoms that have lost one or more electrons) flow through the *electrolyte* (a liquid, paste, or film that allows the ions to pass through) to reach the cathode. The positive ions then combine with electrons and atoms on the cathode to form molecules.

When enough ions on the anode have moved to the cathode, the battery is discharged. When this happens to your laptop or car, for example, you have to charge the battery. Nano has some tricks up its sleeve that could help you avoid those "darn, the battery died" moments. Note that the types of batteries where nano is making a difference are the ones in cars, laptops, and industrial power backups, rather than the batteries rolling around in your kitchen drawer.

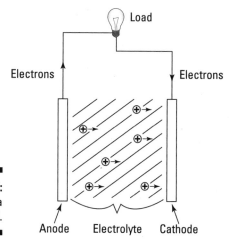

Figure 10-6:
Inside a
battery cell.

Making batteries more efficient

So how does nano help batteries work more efficiently? A key factor in battery efficiency is the *power density,* or how much electrical power a battery can supply per weight. Of the various types of batteries in use, lithium-ion batteries have the greatest power density, which is why they're used in laptops and electric cars.

Lithium-ion batteries can store the same amount of power as a nickel-metal hydride battery in a lighter and smaller package. Lithium-ion manufacturers project that their batteries will last about ten years, about four years longer than nickel-metal hydride batteries.

However, lithium-ion batteries are slower to charge and have safety issues. (For example, in one publicized instance a few years ago, lithium-ion batteries in laptop computers caught fire.) Many companies are exploring the use of nanotechnology to change the material used in lithium-ion battery electrodes. Each company has used its own proprietary material composition, both to reduce the risk of battery fires and to incorporate the capability of a nanostructured surface to increase the surface area on the electrodes. This increase in surface area provides more places for the lithium ions to make contact, allowing greater power density and faster battery discharging and recharging.

We expect the next few steps in nano-based battery improvements to increase power density over conventional lithium-ion batteries by five to ten times. These changes should make laptops lighter, allow them to go a longer time before recharging. These improvements, however, depend on the manufacturers' choices. For example, electric cars may be capable of running a few hundred miles between recharges or may contain smaller, less expensive battery packs.

Upcoming nanotechnology improvements to batteries

Using nanotechnology in the manufacture of batteries could increase the power available from a battery through several methods being developed today. For example:

- ✔ **Coating the electrode's surface with nanoparticles, nanowires, or other nanostructures:** Because the anode is where lithium ions are stored when the battery is charged, increasing the number of stored ions increases the stored electrical power. Various researchers and companies working with lithium-ion batteries have used nanomaterials to develop anodes with a greater density of locations to which lithium ions can attach. This technique increases the surface area of the electrodes, allowing more lithium ions to be stored.

✔ **Changing the atoms to which the lithium bonds:** Changing these atoms changes the electrochemical reaction, which could give off more energy, increasing the power produced by the battery for any given weight.

These techniques could increase the efficiency of hybrid or electric vehicles by significantly reducing the battery weight or increasing their range.

The challenges of nano energy storage

Several groups have demonstrated the potential of batteries with five to ten times the power density of currently available lithium-ion batteries. But don't grab your car keys to go battery shopping. Several issues must be resolved before these techniques are incorporated into batteries in an affordable way.

Researchers at Stanford University have grown silicon nanowires on a stainless steel *substrate* (a structure on which a substance can be grown or deposited). Batteries built with anodes using these silicon nanowires have up to ten times the power density of conventional lithium-ion batteries. The silicon nanowires eliminate the problem of silicon cracking that occurs when using bulk silicon. The cracking is caused by the swelling of silicon as it absorbs lithium ions while the battery is recharged and the contraction of silicon as the battery is discharged (when the lithium ions leave the silicon). Researchers found that the silicon nanowires swell and contract but don't crack.

The busy folks at MIT have developed a technique to deposit aligned carbon nanotubes on a substrate for use as the anode, and possibly the cathode, in a lithium-ion battery. Organic molecules attached to the carbon nanotubes help them align vertically on the substrate. The molecules contain many oxygen atoms that provide points that lithium ions can attach to. This method could increase the power density of lithium-ion batteries significantly, possibly up to ten times. A battery manufacturer called Contour Energy Systems has licensed this technology and is planning to use it in their next generation of lithium-ion batteries.

Another promising way to increase the power density of batteries is to incorporate sulfur in the cathode. The cathode is the electrode that the lithium ions move to when the battery is discharged. In lithium-sulfur batteries, the cathode is a combination of conventional carbon and sulfur. The lithium ions attach to sulfur molecules and the carbon conducts electrons to and from the wires outside the battery. However, sulfur can dissolve from the cathode, limiting the battery's lifetime. Researchers at the University of Waterloo demonstrated that they could significantly reduce the loss of sulfur from the cathode by fabricating the carbon with many tiny nanopores filled with sulfur. Various researchers have been trying for years to get lithium-sulfur batteries to work because the electrochemical reaction with lithium sulfur gives off more energy, which should make the power density of lithium-sulfur batteries about three or four times higher than lithium-ion batteries. The work of these researchers brings us closer to the day when lithium-sulfur batteries will become practical.

Storing electrons with ultracapacitors

Capacitors are like batteries in that they store electrical energy. They simply store electrons on one electrode. Because like charges repel, these electrons push electrons away from any adjacent electrode. These stored electrons can then be used to power electronic gadgets.

One option being explored is making ultracapacitors with nanotubes. Capacitors have one advantage over batteries in that they can be recharged much more quickly. However, capacitors and ultracapacitors do not have as high a power density as batteries. Researchers at MIT are planning to change that by using carbon nanotubes to increase the surface area of the electrodes, thus increasing the number of electrons that can be stored.

The MIT researchers are projecting that they can produce ultracapacitors not only with a power density as great as batteries but also with a much longer lifetime than batteries, up to the lifetime of a car. These ultracapacitors could be recharged in the time it currently takes to fill up your gas tank.

If the power density of ultracapacitors does increase sufficiently to allow them to power electric cars or back up the public electricity grid, their longer lifetime (perhaps decades) would make them a better option than batteries that have to be replaced every few years.

Other Energy Options

When it comes to energy, it isn't all about batteries, solar cells, and fuel cells. Some other interesting ways of conserving energy with nano are being explored today:

  ✔ **Increasing the electricity generated by windmills:** An epoxy containing carbon nanotubes is being used to make stronger and lower-weight windmill blades (see Figure 10-7). The resulting longer blades increase the amount of electricity generated by each windmill.

Figure 10-7:
Windmills
used to
generate
electricity.

Photo from the U.S. Department of Defense

- ✔ **Generating electricity from waste heat:** Researchers have used sheets of nanotubes to build thermocells that generate electricity when the sides of the cells are at different temperatures. These nanotube sheets could be wrapped around hot pipes, such as the exhaust pipe of your car, to generate electricity from heat that is usually wasted.

- ✔ **Clothing that generates electricity:** Researchers have developed piezo-electric nanofibers that are flexible enough to be woven into clothing. When piezoelectric material is compressed or expanded, it generates an electrical current. The fibers can turn normal motion into electricity to power a cell phone and other mobile electronic devices.

✔ **Making biodiesel from algae:** Algae plants are largely composed of oil. Conventional methods to extract the oil crushed the algae, requiring replanting and growth from scratch. Researchers at Iowa State University, the U.S. Department of Energy Ames Laboratory, and a company called Catilinare are developing a method to remove the oil from the algae without destroying the plants, allowing the algae to continue producing oil for another harvest. These researchers use nanoparticles with many pores to soak up the oil like a sponge. This oil can then be processed to make biodiesel.

Because new applications and developments in nanotechnology and energy occur regularly, visit this book's companion web site to find out about any changes: www.understandingnano.com/nanotechnology-energy.html.

Chapter 11

Improving the Environment

*O*ne of our most challenging global problems is our polluted environment. We've managed to contaminate our air and water in many parts of the world through industrial waste, car emissions, oil spills, and more. Nanotechnology holds great promise for helping us clean up the mess we've made and reduce the amount of pollution we generate in the future.

In this chapter we cover ways that nanotechnology is already being used to make our air and water cleaner, and the role it may play in making our future world a better place to live in.

Clearing the Air

Although you can find many opinions about the effect of manmade carbon dioxide emissions on our environment, most scientists believe that the increased level of carbon dioxide in the atmosphere since the industrial revolution is a major cause of global warming.

By understanding some of the causes of pollution in our air, you'll be better able to understand how nanotechnology can help solve the problems we experience today. So, we start by giving you a little background on fossil fuels.

Fossil fuels are formed by the decomposition of organic matter (such as dinosaurs), a process that takes many millions of years. These organisms contain carbon, because all life on our planet is carbon-based. Fossil fuels that contain lots of carbon include coal, petroleum, and natural gas.

When you burn fossil fuels, you produce carbon dioxide. Since the industrial revolution, our species has increased its use of fossil fuels, which has added a lot of carbon dioxide to our atmosphere. In addition, we've systematically hacked away at forests. When we destroy trees, which turn carbon dioxide into oxygen through the process of photosynthesis, we leave more carbon dioxide in our air.

So, what's so bad about carbon dioxide? Carbon dioxide allows light energy from the sun to enter our atmosphere, but it stops some of the energy from the heat on our planet from escaping into space. This retained energy builds up in our atmosphere over time, producing the famous greenhouse effect that many people think is causing catastrophic changes to our planet. Some researchers have predicted that this effect could result in an increase in the earth's temperature of as much as 4.5 degrees Fahrenheit over time. That seemingly small difference would cause glaciers to melt — and if that happens, all kinds of changes to our environment would follow.

In addition to being linked to the greenhouse effect, burning fossil fuels has also been linked to several health problems in humans. Smog generated from cars and trucks and industrial emissions can be a serious health issue, as indicated by this statement from the American Heart Association:

> Fine particulate matter (PM2.5) [has] a causal relationship to cardiovascular disease. . . . The major source of PM2.5 is fossil fuel combustion from industry, traffic, and power generation. Biomass burning, heating, cooking, indoor activities and forest fires may also be relevant sources, particularly in certain regions. Growing evidence also shows that longer-term PM2.5 exposures, such as over a few years, can lead to an even larger increase in these health risks.

Indoor air pollution is another concern. The inside of your house or office building has more organic pollutants than the air outside (2 to 5 times higher, to be specific). This increased air contamination comes from cooking food, as well as from cleaning products, paint, and other everyday items that contain damaging chemicals.

Cleaning the Air with Nanomaterials

Nanotechnology can help clean the air in several ways. It can provide methods of generating energy that don't involve burning fossil fuels. Nano can also help capture carbon dioxide during industrial processes and convert it to fuels such as methane. Nanotechnology could make it possible to make changes to our atmosphere so energy is reflected into space to alleviate the effects of global warming. Nano could even reduce the amount of car emissions released into our air by offering alternative energy sources.

Capturing carbon dioxide with nano

Much of the production of carbon dioxide comes from power plants burning fossil fuels such as coal, oil, or natural gas. According to the U.S. Environmental Protection Agency, "The process of generating electricity is the single largest source of CO_2 emissions in the United States, representing 41 percent of all CO_2 emissions."

Coal burning plants are still the main type of electricity generator in the United States, producing as much as 36 percent of our power.

It is possible to capture carbon dioxide at a fossil fuel–powered power plant and store it, or even turn it into another form of fuel such as methane. The first step in this process is to capture the carbon dioxide produced by power plants. Researchers are developing the following methods:

✔ **Use of nanoporous membranes to remove carbon dioxide from power plant smokestacks:** In a new type of membrane, carbon nanotubes form the nanopores. Carbon dioxide (CO_2) molecules flow through the nanotubes to a storage tank, and the rest of the exhaust stream, largely nitrogen, continues out the smokestack. Carbon nanotubes, unlike other nanopores, have a very smooth inside surface. Therefore, after molecules enter the openings of these nanotubes, they encounter less resistance and move through more efficiently, as illustrated in Figure 11-1.

Figure 11-1: Molecules flowing through a carbon nanotube to filter out contaminants.

Photo from Lawrence Livermore National Laboratory

✔ **Development of nanomaterials to trap carbon dioxide:** Researchers at UCLA are building structures called metal-organic frameworks (MOFs). These structures take their name from the fact that metal molecules and organic molecules are connected in a framework. Researchers have designed MOFs with pores that are just the right size to let carbon dioxide molecules in. Cavities inside the MOFs provide space to store the carbon dioxide.

In the early 1990s, a researcher named Yaghi invented metal-organic frameworks (MOFs). These so-called crystal sponges contain pores that offer an easy way to store gases that are otherwise hard to store or transport.

After you capture carbon dioxide using a method such as the ones just described, you need to get rid of it (a process that has the fancy techie name of sequestration). Figure 11-2 illustrates the sequestration process with a power plant pumping the captured CO_2 underground to be stored in permeable layers of rock. The carbon dioxide can also be pumped into oil fields to boost the recovery of oil by increasing the pressure in the oil field and reducing the viscosity of the oil so that it flows more easily. With the carbon dioxide safely tucked into the ground, it can't contaminate the air.

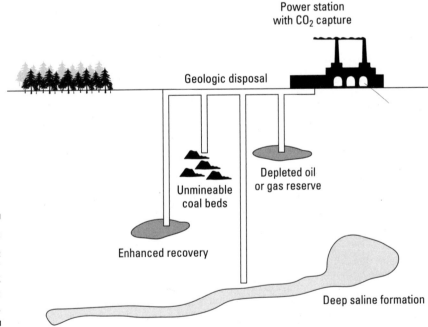

Figure 11-2: Storing CO_2 in permeable rock under a power plant.

Adapted from an image courtesy of the United States Department of Energy

Converting carbon dioxide into something useful

An alternative to sequestration is to convert CO_2 into a combustible fuel such as methane (the natural gas many of us burn in our stoves). All you need to do to make this conversion is to replace the two oxygen atoms with four hydrogen atoms, resulting in CH_4 (methane).

At Penn State, a team of researchers is working on this method of turning captured CO_2 into methane. They use clusters of titanium oxide nanotubes coated with a catalyst that helps convert carbon dioxide and water into methane using sunlight as the power source, as shown in Figure 11-3.

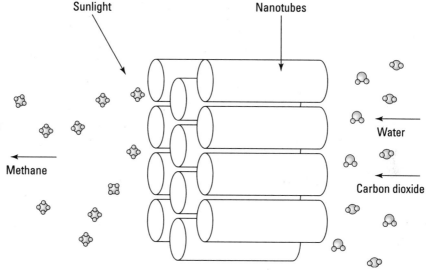

Figure 11-3:
Flowing carbon dioxide through catalyst coated nanotubes to produce methane.

This approach could allow us to create methane from carbon dioxide, store it near a power plant, use the methane to generate more energy, take the carbon dioxide from that process to produce more methane, and so on in a so-called closed loop.

Because this conversion would be powered by sunlight, the additional energy costs in turning carbon dioxide back to fuel to power a plant shouldn't be significant. Of course, there would be capital costs to install the photocatalytic cells and the piping to deliver the carbon dioxide that was separated from the rest of the exhaust stream.

According to Craig Grimes, the leader of the Penn State research group, this method might be used more broadly in the future by capturing and reusing the CO_2 in our vehicles so that none of it is released into the atmosphere.

Another alternative for dealing with the problem of excess carbon dioxide is to generate electricity with a method that does not generate carbon dioxide, such as solar power or windmills. Nanotechnology is being used to make these methods more cost effective when compared to electricity produced by fossil fuel. See Chapter 10 for more about how nanotechnology is being used to generate electricity more cleanly.

Using nanoparticles to fix global warming

Scientists in a field called geoengineering are investigating ways to counter the global warming attributed to high levels of carbon dioxide in our atmosphere. Volcanoes have provided these scientists with a great example of one way to cool the earth.

When a volcano erupts, it sends clouds of particles and gasses into the atmosphere. These clouds contain sulfur dioxide, which can rise as high as the stratosphere. At that height, the sulfur dioxide combines with water vapor and produces sulfuric acid aerosols that reflect the sun's energy, thereby reducing the amount of heat that gets through to our atmosphere.

Monkeying with nature

With the method outlined in this section, it should be possible to control the temperature of our planet. But this possibility raises some important questions:

✔ What temperature do we set? If you've ever fought with your spouse or children about the setting on your living room thermostat, you'll understand this one immediately. Do people living in Siberia get to raise the temperature so they can sunbathe (or at least develop agriculture)? Should folks in the Maldives get a say, because global warming will someday cause the sea to rise and engulf their islands? Or should desert dwellers get some relief?

✔ What is the chance of geoengineers making a mistake? How long would it take to correct that mistake (if we can), and how much damage could occur?

✔ Will the effort to reduce carbon dioxide emissions be too slow? Given the economic and political realities of reaching agreement on environmental issues, as well as the time it takes for carbon dioxide levels in the atmosphere to respond to reduced levels of CO_2, will our efforts come too late?

If the worst-case predictions of coastal areas flooding from rising ocean levels caused by melting ice caps are valid, there may be no choice but to put a geoengineering fix in place as a patch until carbon dioxide levels in the atmosphere are reduced. However, we shouldn't depend on geoengineering and nano to ride to the rescue in time. At some point, no amount of geoengineering will save us, so just to be safe, you might consider walking or riding a bike to work today.

When the sulfuric acid returns to earth in rain, the side effect is acid rain. There's always something. . . .

The resulting lowering of the atmosphere's temperature can seem small, but it can be significant in terms of its effect on our environment. For example, the U.S. Geological Survey estimates that an eruption of Mount Pinatubo in the Philippines in 1991 sent about 20 million metric tons of sulfur dioxide into the atmosphere and caused about half a degree centigrade (about 1 degree Fahrenheit) cooling in the northern hemisphere.

Here's where nano comes in. A researcher at the University of Calgary has designed particles composed of different nanofilms that could be released into the atmosphere to cool the earth without some of the negative effects caused by volcanoes. The top layer of a nanofilm protects the middle layer from oxidizing; the middle layer reflects light; and the bottom layer interacts with the atmosphere's electric field to orient the disk-shaped particle horizontally for optimum reflection. That reflection cuts down the amount of sunlight that reaches our atmosphere and helps cool our planet slightly to compensate for global warming.

Making cars cleaner

This may not come as a big surprise to you, but air pollution from cars and trucks can be a threat to your health, as indicated by this statement from the American Heart Association: "The scientific evidence linking air pollution to heart attacks, strokes and cardiovascular death has 'substantially strengthened,' and people, particularly those at high cardiovascular risk, should limit their exposure."

A very fine particle called PM2.5 seems to be one of the most important links between air pollution and cardiovascular disease. The major source for PM2.5 is fossil fuel emissions from cars, power plants, and other forms of combustion such as forest fires. The longer the exposure to PM2.5, the worse the health risks, which include inflammation and irritation of nerves in lungs and inhibited circulation. Inhaling PM2.5 can cause heart attacks, strokes, and in some cases, death.

To avoid exposure to PM2.5 you can stay inside and out of traffic on days with high pollution levels, wait for our governments to clean up the air by passing legislation, or move to the country. As none of these options is practical, you might take heart at the promise nanotechnology holds for reducing the amount of fossil fuels that we produce with vehicles.

Nanotechnology can help provide alternatives to burning fossil fuels to run our cars in a few ways:

- ✔ **Using nanomaterials in hybrid car batteries:** This solution would increase the number of ions that can attach to the electrodes in the battery, therefore increasing the power density of the battery. This new type of battery should improve the performance and reduce the price of lower-pollution plug-in hybrid vehicles, as well as pollution-free plug-in electric vehicles.

- ✔ **Using nanotubes in ultracapacitors:** This method would increase the number of electrons that can be stored in a capacitor, which should increase the amount of electrical power available from ultracapacitors to about the same as similarly sized batteries. Because capacitors can be recharged much more quickly than batteries, using ultracapacitors in plug-in electric vehicles would shorten the time needed to recharge, making the recharge process as quick as filling a gas tank.

- ✔ **Using nanomaterials to make fuel cells more practical:** Nanomaterials are being used to improve the efficiency of the catalyst used in fuel cells, which allows less platinum to be used, thus lowering the cost of fuel cells. Also, graphene is showing promise as a storage medium for hydrogen and could be used for lightweight hydrogen fuel tanks in cars powered by pollution-free hydrogen fuel cells.

We discuss how nano improves batteries, ultracapacitors, and fuel cells in detail in Chapter 10.

Cleaning the air in your house

It's not just the air around the nearest highway or near your local industrial plant that's causing problems in our environment. According to the U.S. Environmental Protection Agency, the organic chemicals in paint, varnish, various cleaning solutions, cosmetics, wax, and other everyday products can release organic compounds called VOCs (volatile organic compounds). VOCs can irritate your eyes, nose, and throat. You can get headaches or grow nauseous if you're overexposed, and there could even be damage to your liver, nerves, and kidneys. VOCs have also been linked to cancer.

Researchers in Japan have created a new material that removes VOCs, nitrogen oxides, and sulfur oxides from room-temperature air. To do this, they use gold particles embedded in a very porous manganese oxide. This catalyst scrubs pollutants out of the air much more effectively than any method used to date.

The manganese oxide contains nanoscale pores, which provide lots more surface area. That greater surface area provides more points where volatile molecules can be absorbed. The pollutants that are absorbed can then be broken down through a reaction among the VOCs, gold nanoparticles, and manganese oxide.

Cleaning Water

The air we breathe is pretty darn important in the scheme of things, but the water we drink is also right up there in terms of our survival. Our actions, from releasing chemicals into our groundwater to oil spills can cause serious damage to this precious resource. Nanotechnology may once again hold the key, this time to cleaning our world's water supply.

Understanding the problem

The United Nations has published figures that show that in the year 2000, more than two million people died from drinking contaminated water. What that figure doesn't address is the damage that water pollution causes to animals and marine life, and the fact that most of our water (that is, saltwater) isn't drinkable at all.

Water pollution can be caused by several things, for example:

- Heavy metals produced in industrial processes
- Industrial solvents
- Contaminants from sewage
- Sulfate particles in acid rain

In addition, the balance of life in bodies of water can be thrown off by particles in water that prevent sunlight coming through, killing plants and organisms that are vital to marine life.

The U.S. Environmental Protection Agency has identified more than 1,200 sites that are known to have released or are threatening release of hazardous substances. These hazardous substances range from organic cleaning solvents to radioactive particles. One concern about such sites is that substances could seep into groundwater and contaminate sources of drinking water. Figure 11-4 shows some of the ways that our activities can produce pollutants that leak into groundwater.

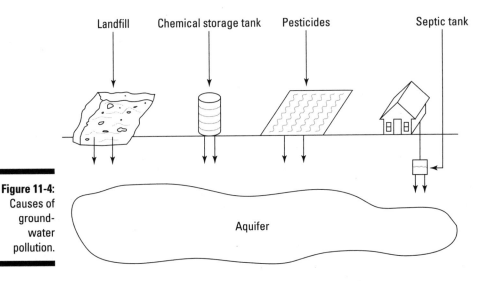

Landfill   Chemical storage tank   Pesticides   Septic tank

Figure 11-4:
Causes of
ground-
water
pollution.

Aquifer

Nanotechnology can help clean water using several methods, including injecting nanoparticles underground to clean contaminates out of groundwater, using nanowires to build a membrane that can soak up oil spills, and using silver nanowires and carbon nanotubes to kill bacteria in drinking water.

In addition, nanomaterials can make desalination of seawater more efficient and cost effective to bring potable drinking water to millions of people who live near our oceans.

Cleaning contaminated groundwater with nano

Industrial solvents in groundwater, such as a liquid cleaning solvent called trichloroethylene (TCE), can damage our health. According to the Agency for Toxic Substances and Disease Registry, a department of the U.S. Department of Health and Human Services, "Drinking or breathing high levels of trichloroethylene may cause nervous system effects, liver and lung damage, abnormal heartbeat, coma, and possibly death. Trichloroethylene has been found in at least 852 of the 1,200+ National Priorities List sites identified by the Environmental Protection Agency (EPA)."

One challenge in cleaning groundwater is the removal of industrial water pollution, such as TCE. Researchers have shown that iron nanoparticles can rapidly degrade chlorinated hydrocarbons, such as TCE, to make a harmless mix of byproducts. When you inject iron nanoparticles into a groundwater system such as an aquifer (permeable rock that contains groundwater), a reaction occurs between the iron atoms, the solvent, and hydrogen ions. In this reaction, iron atoms from the nanoparticle give up two electrons (iron atoms can easily give up electrons and are quite stable as positive ions in water) to the chlorine atoms. As a result, hydrogen ions (occurring naturally in water) combine with the carbon atoms. This leaves you with a common, ordinary organic molecule in which the carbon atoms are surrounded by hydrogen atoms (like methane or ethane), positive iron ions, and negative chlorine ions (commonly present in saltwater). Because nanoparticles can remain suspended in the groundwater for a long time, they can permeate large areas of groundwater.

The iron used in these nanoparticles is often referred to as zero-valent iron (ZVI). That's just a fancy way for chemists to say that the iron hasn't reacted with any other material yet and therefore still has all its electrons.

Using iron nanoparticles in this way to treat groundwater, as illustrated in Figure 11-5, is much less expensive than pumping the water out of the ground for treatment.

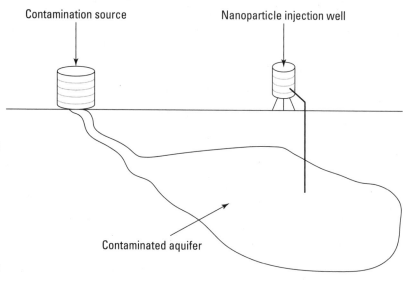

Contamination source

Nanoparticle injection well

Figure 11-5:
Injecting nanoparticles into a contaminated aquifer.

Contaminated aquifer

Lehigh Nanotech is a company that is using iron nanoparticles to clean contaminates. The intriguing techniques they are using were developed through research at Lehigh University. You can find out more about their activities at www.lehighnanotech.com.

Metallic contaminates in groundwater, such as mercury, are also a concern. Researchers at Pacific Northwestern Laboratory have developed a material to remove mercury from groundwater. The material is called SAMMS, which is short for Self-Assembled Monolayers on Mesoporous Supports. This material consists of ceramic particles whose surface has many nanosize pores. These pores are lined with molecules that have sulfur atoms on one end, leaving a hole in the center that is also lined with sulfur atoms. Scientists line the nanopores with molecules containing sulfur because it bonds to mercury, so mercury atoms bond to the sulfur and are trapped in the nanopores. The researchers are developing SAMMS that contain other molecules in the nanopores to remove other metal contaminates from groundwater.

Cleaning oil spills

Oil spills, like the one involving British Petroleum in 2010, can be devastating to our environment, costly, and deadly to marine life and birds. Nanotechnology may hold some promise for helping clean up such spills more quickly and efficiently.

Researchers at MIT report that they have developed a nanomaterial that can be used to absorb as much as 20 times its weight in oil. According to Francesco Stellacci, who is leading the research team, "What we found is that we can make 'paper' from an interwoven mesh of nanowires (a mat) that is able to selectively absorb hydrophobic liquids — oil-like liquids — from water." In addition, the nanowire mat can be recycled and used several times.

Folks at MIT are also working on robots that could troll the surface of water, gather oil in this nanowire mat, and process it on the spot. Called Seaswarm, this robot system will act like a swarm of bees working to suck up oil — without human involvement. The robots will actually be able to communicate wirelessly and use GPS to monitor their own location so they can spread out evenly over a spill site. The swarm detects the edge of the spill area, and works inward.

Making saltwater drinkable

While there is lots of water around, only a small portion of it is drinkable. According to the USGS (United States Geological Survey), our oceans contain about 96.5 percent of all the water on Earth, but obviously we can't drink saltwater. Then there are problems in the supply and quality of water. The problems range from the population of cities being too large for the amount of water available in rivers and lakes, to rural villages where the available water is polluted but they have no water treatment facilities. (The United Nations estimates that 1.1 billion people do not have access to water treatment facilities.)

Given these problems, the idea of making saltwater drinkable is appealing. However, the cost of doing so is currently very high. For communities near seawater, the good news is that nanotechnology is helping to make desalination of saltwater more economical.

The natural movement of water molecules into saltwater to equalize the concentration of contaminates is called osmosis. The process of forcing water molecules from saltwater to freshwater is the reverse of this process and therefore is called reverse osmosis.

The reverse osmosis process, shown in Figure 11-6, is used in many desalination plants. In this system, you place saltwater under pressure, which forces it through a membrane. The membrane is generally a polymer with many nanosize pores, and is a material that allows water through the pores but stops salts and bacteria.

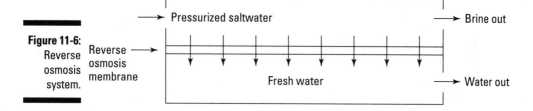

Figure 11-6: Reverse osmosis system.

Reverse osmosis systems require pumps to maintain sufficient pressure to force the water through the membrane as well as cleaning procedures to clean bacteria that grows on the saltwater side of the membrane, referred to as fouling.

Researchers are investigating the use of nanomaterials to reduce the pressure needed to force water through the membrane and to reduce the capability of bacteria to grow on the membrane.

Making water flow

We may be able to benefit from carbon nanotubes in the membranes that are used in reverse osmosis to help with the process of desalination. For example, a company called NanOasis is working on membranes that contain a very dense polymer film with carbon nanotube pores. Because the inside of carbon nanotubes is very smooth, water is transported through them more easily. And, while the nano pores allow water to flow through, they stop salt ions, making this method perfect for desalination. This method could reduce the energy required for desalination by 30 to 50 percent.

Reducing bacteria in reverse osmosis

A company called NanoH$_2$O adds nanoparticles to their membrane to optimize properties such as surface roughness and charge. The company has been able to reduce the chance of bacteria adhering to the membrane. Because bacteria on the membrane can reduce the amount of water passing through, reducing the bacteria on the membrane means that you don't have to shut down the system for cleaning as often.

Exploring capacitive deionization

A desalinization method called capacitive deionization has the potential to become more cost effective than reverse osmosis. As shown in Figure 11-7, a capacitive deionization cell contains two electrodes, one positively charged and one negatively charged. The electrodes are charged because salt consists of negative and positive ions. Because opposite charges attract, negatively charged ions are attracted to the positively charged electrode and positively charged ions are attracted to the negatively charged electrode. For example, if you dissolve regular table salt (sodium chloride) in water, the sodium and chloride separate to form positively charged sodium ions and negatively charged chloride ions.

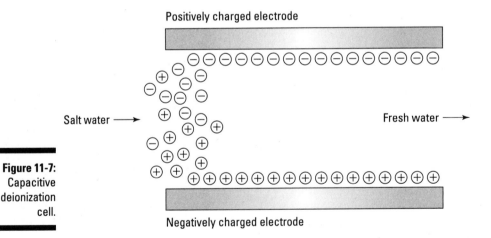

Figure 11-7: Capacitive deionization cell.

As seawater runs through the cell, the salt ions attach to the electrodes and deionized water leaves the other end of the cell. This technique doesn't require high pressure to push the water through the membrane, as in reverse osmosis, so it would be less expensive and use less power.

Researchers are developing electrodes made with nanomaterials to increase the electrode surface area, which should increase the speed at which a cell can remove salt ions from seawater. One interesting technique is the use of electrodes constructed from graphene flakes, which researchers at the University of South Australia have demonstrated. Researchers from around the world are also attempting to develop low-cost capacitive deionization systems using nanostructured electrodes.

Removing bacteria from drinking water

Nanotechnology is being used to develop low-cost filtering technology for communities near contaminated water sources.

Many water sources contain bacteria that cause diseases ranging from giardiasis, familiar to hikers, to cholera. While municipal water systems have methods to kill bacteria, many towns in developing countries don't have municipal water systems.

Removing arsenic from well water

A problem for tens of millions of people around the world is the presence of arsenic, a naturally occurring substance in soil that can dissolve in water, including well water. Robert Bunsen (the developer of the Bunsen burner, which you might remember from high school science labs) determined in the 1830s that when you mix ferric oxide with arsenic, you get a mixture that the fluids in your body and water cannot break down. Based on Bunsen's discovery, scientists started using iron oxide in filters to remove arsenic from water; however, the cost of this process is out of reach for many people in the third world. Researchers are working on the use of iron oxide nanoparticles to increase the surface area of iron oxide available to react with the arsenic, thereby reducing the cost of the material you need to make it work.

Although there are arsenic-contaminated wells in many countries, including the United States, more developed countries have other water sources. However, a low-cost method for removing arsenic from water could be essential in less developed countries such as Bangladesh. In many of these countries, millions or tens of millions of people with very little income drink arsenic-contaminated well water daily. Ironically, these wells were drilled so that people could avoid drinking from contaminated surface water sources.

A key aspect of research efforts is to see how theories developed in the lab will work in the field. For example, researchers at Rice University are currently working to develop a low-cost method of using iron oxide nanoparticles. These researchers are currently performing field evaluations of the iron oxide nanoparticle method with arsenic-contaminated water in wells at Guanajuato, Mexico.

Researchers at Stanford University have developed a low-cost water filter that combines silver nanowires, carbon nanotubes, and electricity to kill these bacteria. Conventional filters try to block bacteria and allow the clean water to flow through, but this method requires very small holes in the filter through which the water has to be pumped. Nanotechnologists at Stanford figured out a way to make the holes in this filter large enough so that water would flow through without pumping.

However, bacteria can still pass through the filter. To get rid of most of the bacteria, they formed the filter material by dipping cotton fabric in a solution containing silver nanowires and carbon nanotubes. The cotton is coated with the nanotubes and nanowire, which allows it to conduct electricity. By applying 20 volts across the conductive filter, they can kill 98 percent of the bacteria.

Researchers think using multiple stages of the filter will get rid of the last few percent of bacteria. The silver in the nanowires will help prevent the growth of any bacteria on the filter, thus preventing fouling of the filter. This method is designed to be a low-cost, low-maintenance, and low-energy way to filter water for use in developing countries.

Chapter 12

Star Wars: Nano in Space and Defense

*A*s our governments go about the business of spreading our reach in space and defending themselves, nanotechnology is one of their most important partners. That's because nanotechnology can make materials lighter, stronger, more flexible, and able to run with less energy.

These characteristics have many applications in space and in battle. In this chapter, we cover exciting advances in space, such as a space elevator to lift people and cargo into space, advances in sensors to explore new worlds, and spacesuits designed to keep astronauts healthy. In the military realm, lightweight armor and fabrics that defeat chemical weapons are on the horizon, as well as remarkable air and ground craft that can literally change their shape and, in some cases, their function.

Space: The Final Frontier

Travelling in space offers a unique environment where the usual rules don't always apply. Getting spacecraft into space, getting supplies up to space stations, and providing materials and equipment that work in cramped quarters are all challenges that nanotechnology can help address.

Boosting space travel with lightweight spacecraft

When it comes to the cost of travelling from A to B, you're not the only one watching your pennies. One challenge to space travel today is the cost of fuel to send a rocket into orbit. By making spacecrafts lighter, we can minimize the amount of fuel needed to send them into space. Nanotechnologists estimate that using high-strength, lightweight materials could reduce the weight of spacecraft by as much as 30 percent. Researchers at NASA's Glen Research Center are developing composites of polymer and nanoparticles that provide such high-strength, lightweight materials. One of the nanoparticles they're working with is carbon nanotubes. These nanotubes are *functionalized* (the properties of the nanotubes are customized to perform a specific function) by bonding oxygen atoms to their surface. The oxygen atoms help the carbon nanotubes couple with the polymer matrix, which provides the composite with greater strength.

These researchers are also working to strengthen materials using silica aerogels. An aerogel is a gel in which liquid is replaced by gas, making it a great insulator. Silica aerogels are composed of silica (silicon dioxide, the same stuff that makes up glass) nanoparticles interspersed with nanopores filled with air. Because of all that air, nano aerogel is one of the best thermal insulators known to man.

Researchers have found that coating the surface of silica aerogel with polymers increases the strength so much that these gels can actually support a mechanical load. This characteristic could allow an aerogel material to serve several functions in cryogenic fuel tanks, providing both insulation and additional strength.

Aerogel was created by Samuel Stephens Kisler in 1931. He made a bet with a buddy about who could replace the liquid in jelly with gas first, without causing the jelly to shrink. Kisler won the bet. His method produced a bluish haze, so this material, which feels a bit like Styrofoam, has been referred to as blue smoke.

Composites and silica aerogel are just the tip of the iceberg when it comes to how nanotechnology could change the way that spaceships are made. NASA has included a concept called self-healing spaceships in their 2030 nanotechnology roadmap. Just as your skin heals a small puncture wound, the folks at NASA are hoping that nanotechnology can provide a way for the skin and structural components of a spaceship to seal damage from meteors and other debris.

NASA is also planning to use nanosensors to improve the monitoring of spaceship systems such as life support. The capability of nanosensors to quickly report changed levels of trace chemicals in the air could be very useful to keeping life support systems working correctly in a spaceship's closed system. A longer-term proposal is to place nanosensors throughout the skin of a spacecraft to act like the nerve endings in your skin. When a particular region of the spacecraft skin becomes stressed or damaged, an alert is sent to the main computer to take action and alter the spaceship's course, just as you would jerk your hand away from a hot stove.

Taking off with the space elevator

Another challenge we have to solve before we can roam around space like Captain Kirk is how to get supplies and people into space affordably. Currently this involves sending expensive spacecraft out there like some extraterrestrial grocery delivery service.

The space elevator is a device that could provide an alternative way to put things in orbit. Just like the elevator in an office building, the space elevator, illustrated in Figure 12-1, could be used to transport both materials and people. As with conventional elevators, the space elevator will be fitted with a cable. However that cable will have to be stronger than any cable you've ever seen. Roughly 90,000 kilometers long, the space elevator cable will be anchored at the top to an asteroid (called the counterweight) in orbit around the earth, and at the bottom by an anchor station, perhaps floating in the ocean (similar to an oil-drilling rig). Such a cable will probably be made from carbon nanotubes.

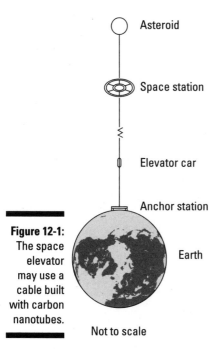

Asteroid

Space station

Elevator car

Anchor station

Earth

Figure 12-1: The space elevator may use a cable built with carbon nanotubes.

Not to scale

Researchers at the Macromolecular Materials Laboratory at Cambridge University have developed a method to make carbon nanotube fibers more cheaply by spinning them into a fiber as they come out of the vapor deposition reactor (a machine filled with hot carbon gas) where they are created. The researchers have managed to produce samples several times stronger than steel and are working to increase the strength of the fibers, as well as making the quality of the fibers more consistent. They believe that the process could produce nanotube fibers inexpensively when scaled up to production volumes. This and other methods to produce fibers and cables from nanotubes will certainly be used for other applications, such as bulletproof vests, before producing cables strong and long enough for the space elevator.

The space elevator could save us a ton of rocket fuel and make sending cargo into space much more practical. Solar cells would be placed on the space elevator cars. By shining lasers from the anchor or space station onto the solar cells, the system would receive the power required to drive a car up or down the cable. This energy system would reduce the weight that would have been taken up by storing fuel in the car, leaving more capacity for cargo.

Although there are some engineering challenges, perhaps the most intriguing of which is actually stringing this 90,000-kilometer cable between the anchor station in the ocean and the counterweight asteroid in orbit, steps are underway to address those challenges. For example, yearly competitions conducted by the Spaceward Foundation, whose web site is shown in Figure 12-2, are providing a focus for energetic minds to demonstrate prototypes and earn some substantial cash prizes. The competition focuses on the two main challenges of developing the space elevator: One is on developing a cable with the combination of high strength and light weight; the other challenge is to develop climbers powered by lasers on the ground that can climb the cable.

Elevator 2010

The folks at the Spaceward Foundation liken their space elevator competition to various other engineering competitions that help gain public mindshare for this amazing effort. They've drawn comparisons to a time when early aviators were trying to convince the public that there was a future for air travel. In those days, air shows entertained but also conveyed the message that air travel was a viable industry of the future.

In a similar way the Elevator competitions build enthusiasm for the space elevator's future and generate great ideas from a variety of sources. Prize money for the competition is provided by a program at NASA called Centennial Challenges.

Follow the games on the Space Elevator Games Offical Web Site

www.SpaceElevatorGames.org

- The level 1 (2 m/s) challenge was met by team LaserMotive from Seattle, who took home $900,000.
- The level 2 (5 m/s) challenge remains unclaimed.
- All three teams think they can claim the level 2 prize. **This time, it's personal!**

Figure 12-2:
The Spaceward Foundation runs yearly Space Elevator Games.

A report by NASA's Institute for Advanced Concepts gives a good introduction to the techniques necessary to construct the space elevator. You can read the space elevator report at

www.niac.usra.edu/files/studies/final_report/521Edwards.pdf.

The Space Elevator Group has an FAQ page that provides answers to some key questions about the space elevator. Visit them at www.liftport.com.

Letting spacesuits fix themselves

If you're the type who uses the sanitary hand wipes at the grocery store to avoid germs on grocery carts, you will totally relate to the concerns an astronaut might have about picking up a space bug. Researchers at Northeastern and Rutgers Universities have come up with a way to protect astronauts with several layers of bio-nanorobots. In these nano-enabled spacesuits, an outer layer would contain bio-nanorobots. These robots would be able to deal with medical emergencies by administering drugs if the suit wearer were injured or became sick.

The term *bionanorobot* refers to the fact that biological molecules are actually a part of the nanorobot mechanism. We all have proteins in our bodies that have a built-in ability to travel. By using carbon nanotubes to link these proteins to other molecules that make up the nanorobot, we give the robot the capability to hitchhike, wandering around a spacesuit to find problems and heal them.

Keeping astronauts healthy with nanosensors

Nanosensors, which might detect biological or chemical molecules, send us information about the nanoscale world. These sensors have several uses, one of which is in healthcare. Companies are working alongside NASA to develop nanosensors that can analyze the state of your entire body from a single drop of your blood. The plan is to be able to assess immune function, heart health, bone density, the condition of the liver, vitamin levels, and the status of lipids, such as fats, and triglycerides from this one drop of blood.

Eventually an entire hospital lab's capabilities could be found in a portable medical field kit to help assess patients using this testing method virtually anywhere. For astronauts, tests that can now be performed only using large machines in a lab will one day be available in the confined area of a spacecraft.

Finding water (and more) on other planets

Another type of sensor being developed is a biosensor for finding pathogens, such as *E. coli*, in water. The biosensor is a good fit for use in space because the biosensor is small and doesn't require energy to work nor expertise to use it.

Developed by Ames Research Center for Nanotechnology, the sensor is made with carbon nanotubes. Researchers are using nanotubes because they are both good conductors of electricity and will attach to biological molecules. A strand of nucleic acid from a waterborne pathogen is attached to the tip of each nanotube in a sensor. When the sensor is placed in water and one of the strands of nucleic acids on a nanotube encounters a matching strand, the strands bind, forming a double helix. This process releases a small electrical charge. Because the nanotubes conduct electricity, they then send the charge to a transducer located in the sensor. The transducer converts that energy to produce a signal, indicating to an observer that the pathogen has been located.

Walking sensors

Researchers are exploring the use of nanosensors, such as the one described in the preceding section, in robots such as the TETwalker, shown in Figure 12-3. These robots can be very small and grouped to create *autonomous nanotechnology swarms,* or ANTS. ANTS could change their shape and move over uneven ground or even form themselves into solar sails. (See the next section for more about solar sails.) One scenario could involve a swarm of robots sensing something of interest on another planet and then forming themselves into an antenna to communicate the finding back to Earth.

Courtesy of NASA

TETwalker stands for tetrahedral walker. The TETwalker looks like a tetrahedron (a pyramid with three sides and a base).

To make the robots in swarms smaller, researchers are exploring the use of nano electromechanical systems, or NEMS (which we discuss in more detail in Chapter 4), instead of motors. Using nanotubes helps make the robots not only smaller but also more flexible. Because struts made of metal tape and nanotubes are retractable, the robot can shrink until all its nodes touch.

Researchers are also exploring the use of artificial intelligence to help the robotic nanotechnology swarms move around and work together with other swarms to essentially make decisions about a swarm's operation. The system could learn about and adapt to its environment, helping it to survive and to provide us with better data about other planets.

Currently, if a space robot land rover falls over, it can't get up, so its usefulness is pretty much at an end. TETwalkers move by falling over.

Sailing through space with lightweight solar sails

After you have people and cargo in orbit, you can use nanotechnology to reduce the rocket fuel needed to travel to the moon or planets. Just as sailboats are propelled by wind while on the seas, spaceships can be propelled through space by light from the sun reflected off solar sails. Use of solar sails could mean that the only fuel required would be during liftoff, docking, and landing.

However, solar sails will have to be very large, spreading out for kilometers, and very thin to keep their weight low. That's where nanotechnology enters the picture. Folks at the University of Texas have used carbon nanotubes to make thin, lightweight sheets that may replace the polymer sheets that researchers have experimented with to date. At this point, NASA has built a small solar sail to test the mechanism for unfurling the sail in orbit. Although details still need to be worked out (such as how to unfurl a thin, fragile sail in orbit, along with the continual struggle to reduce weight), this method has great potential for reducing the amount of fuel needed to travel between planets.

Giving spacecraft smaller rockets

Good things come in small packages, or so the people involved in the space program believe. These folks are using small spacecraft satellites more and more because they cost less and can be deployed quickly. This size of spacecraft requires in-space propulsion systems that don't take up a lot of room or weigh much. One solution is called the Nanoparticle Field Extraction Thruster (nanoFET). In this device, nanoparticles are charged by losing electrons when they touch an electrode at a positive voltage. After the nanoparticles are charged, an electric field can accelerate them, providing thrust to the spacecraft.

NanoFets may also be useful for spacecraft that are designed to explore other planets, such as the Galileo (shown in Figure 12-4). Besides their small size, nanoFETs have an additional advantage in space exploration: You can vary the amount of thrust generated over a wide range by changing the voltage used to accelerate the charged nanoparticles. This capability is useful because a spacecraft may need higher thrust when nearing a planet — to put itself into orbit about the planet or to change its orbit — than it would need while cruising between planets. NanoFETs could supply the required range of thrust without the need to build more than one type of propulsion system into the spacecraft.

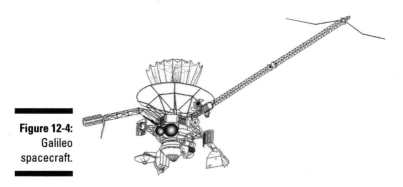

Figure 12-4:
Galileo
spacecraft.

Courtesy of NASA

Drafting Nano for Defense

Military types haven't failed to notice that nanotechnology could make a big difference to how troops function, travel, and stay safe. The options range from what the well-dressed soldier will wear to methods for making aircraft wings that can fly like a bird.

Decking out troops in lightweight body armor

The Institute for Soldier Nanotechnologies (ISN) was established at MIT in 2002. Currently in the second of two five-year contracts with the U.S. Army Research Office, they are in the business of developing and taking advantage of nanotechnology to help soldiers survive in battle conditions.

One way they are trying to do that is by helping soldiers ditch their hefty packs (sometimes more than 140 pounds) in favor of lighter weight materials. Currently, soldiers carry around lots of equipment with no compatibility among the devices. Their clothing also doesn't protect soldiers from bullets to a great enough degree. A nanobattlesuit is being developed that could be as thin as spandex and contain health monitors and communications equipment. Nanomaterials can also provide strength that far surpasses currently available materials, providing bullet shielding that's much more effective. These jump-suit style outfits might even be able to react to and stop biological and chemical attacks. This protection and these devices would be integrated into one suit that would be more efficient and lightweight than current packs.

Nanotechnology also offers the military the ability to miniaturize, which also cuts weight. A weighty field radio could become a device about the size of a button, worn on the collar, for example.

The U.S. Army Natick Soldier Systems Center has published a white paper that discusses how nanotechnology may be used in soldiers' equipment in the future. Visit this URL to read the paper: `http://nsrdec.natick.army.mil/FSI/docs/FSI_WhitePaper_Dec2009.pdf`.

Morphing for higher efficiency

The idea of a piece of equipment that can morph (change its shape) isn't science fiction. Today, research is well under way to develop aircraft wings, propellers, and transport vehicles that can literally change shape to improve their performance and efficiency.

Helping aircraft flap their wings

Researchers have worked on aircraft that swing their wings in close for high-speed flight and extend their wings to provide more lift for takeoff and landing, as illustrated in Figure 12-5. Unfortunately, the hinges that allow the wings to swing add weight, so researchers are developing materials that will need only an electrical voltage to change the shape of aircraft wings and other structures. NASA has developed a carbon nanotube polymer composite that bends when a voltage is applied. They envision that this type of morphing material will be used in various ways.

Figure 12-5:
Two states of a morphing swing-wing aircraft.

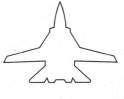

Wings extended

Wings in

So what other advances will the airplane of the future have? Aircraft wings will contain smart materials that make them more aerodynamic and more efficient to control. Such a craft would sense conditions while in flight. Sensors in the wings will measure the pressure on each wing's surface. Using actuators (devices that control a system), the wing can then respond, even changing shape, just as a bird's wing responds to air pressure or weather.

By using such a system of sensors and actuators, along with efficient micro-processors and controls, these aircraft could keep track of not only their environment but also their performance and even the condition of those operating them. This kind of system could save both fuel and lives.

An illustration from NASA, shown in Figure 12-6, gives you an idea of what a future morphing aircraft might look like.

NASA Dryden Flight Research Center Photo Collection
http://www.dfrc.nasa.gov/gallery/photo/index.html
NASA Photo: ED01-0348-1   Date: 2001   Photo by: NASA
An artist's rendering of the 21st Century Aerospace Vehicle, sometimes nicknamed the Morphing Airplane, shows advanced concepts NASA envisions for an aircraft of the future.

Figure 12-6:
A look at a possible future morphing aircraft.

Courtesy of NASA

The military establishment is looking long and hard at morphing aircraft. One reason for their interest is that military aircraft today are designed to perform one kind of mission. For example, one aircraft might excel at recon-naissance, while another is designed for bombing missions. Features such as the capability to carry more weight, high speed, and a small turning radius are difficult to combine in one aircraft. Because their designs are so specific, aircraft can't perform more than one role, and in many cases they are limited to certain airfields or ships to use for takeoff and landing.

But imagine if an aircraft could change its shape. The Morphing Aircraft Structures (MAS) was started to design and build these shape-changing air-craft for the military. If the military is successful, that would mean that one aircraft might be able to perform more than one role and be able to take off and land from more types of airfields or ships. This flexibility could result in huge savings in cost and much improved efficiency.

Making propellers more efficient

DARPA also funds the Mission Adaptive Rotor (MAR) program, which is focused on improving the performance of helicopter rotors (the part that whirs around to lift helicopters into the air). Rotors that can morph would last longer and offer improved performance. These improvements come in part from a reduction in rotor vibration. The improved performance involves an increase in the amount of weight that the helicopter can carry and an extension of its range.

Shape-changing vehicles

Shape changing isn't limited to the skies. The Transformer (TX) vehicle being developed by DARPA can travel on roads but is also capable of vertical take-off and landing. The combo land and air vehicle could be operated by any-body (no pilot's license required) for scouting missions or transporting of troops or supplies.

The Transformer would be able to carry four people and use fuel-efficient energy storage systems such as ultracapacitors and batteries. The body of the vehicle could morph to grow wings or pull them back in based on whether the vehicle is on land or aloft. As military personnel move around in the TX, they could use the capability to fly to circumvent obstacles, go over rough terrain, and avoid landmines or ambush, while retaining the capability to drive on roads.

This project is in its initial stages, so there are more questions about what the TX will look like than there are answers. Visit this book's companion site, `www.understandingnano.com/defense_nanotechnology.html`, to look for updates on progress as they become available.

Storing more energy with nanoenergetics

Energetic materials aren't running on caffeine; they are materials that contain a large amount of chemical energy. When released, this energy can take the form of an explosion, fireworks, or rocket fuel, for example.

Researchers are working with energetic materials at the nanoscale. As shown in Figure 12-7, nanoparticles have more surface area that is in contact between the particles of the different chemicals that make up an explosive. After a reaction is initiated (that is, the explosion is set off), this greater surface area causes a faster reaction rate, which makes for a more powerful explosion. This work could come in handy in weapons systems that would utilize greater amounts of energy, making them more lethal. By working at the nanoscale, weapons designers can also control the rate at which energy is released by changing the size of the nanoparticles, allowing the designers to customize the explosive for each application. For example, a weapon designed to penetrate into the ground to destroy a bunker may need an explosive with a different reaction rate than a weapon designed to explode and project shrapnel above ground troops.

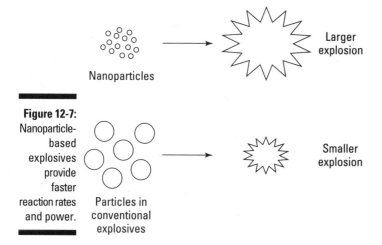

Figure 12-7: Nanoparticle-based explosives provide faster reaction rates and power.

An example of this technology is the use of aluminum nanoparticles in explosives that the Air Force is developing. When you add nano-aluminum powder to explosives, you can make weapons smaller and more powerful. These weapons are useful in aircraft with limited space, such as remote control drones (also called unmanned air vehicles). Researchers are developing techniques that allow weapons manufacturers to add a greater amount of nano-aluminum powder to an explosive using a solvent.

Not only weapons and battle vehicles are benefitting from nanotechnology. Researchers have developed a nanoparticle composite of zinc and carbon that heats itself up by means of electrochemical oxidation when it's exposed to air. This technology is used to heat soldier rations, the so-called meal, ready-to-eat (MRE).

Detecting hazardous agents

Hazardous agents are any chemical or biological materials present in the environment. For soldiers, hazardous agents might come from chemical or biological weapons or an environmental disaster. The Protection and Hazard Mitigation program of the Defense Threat Reduction Agency (DTRA) is focused on possibilities for detecting hazardous agents and alerting troops to their presence.

Nanotechnology is an important part of the "smart" portion of the hazard mitigation system being developed. In this system, a sensor detects an agent and responds by taking action to detoxify or isolate the agent. It then generates a signal to an observer. Nanoscale materials may be used in the decontamination portion of this system in a few ways, working at the molecular level.

Another method of protecting troops from agents uses their own clothing to destroy toxins and prevent the toxins from penetrating the clothing to reach their skin. Using antimicrobial reactive additives as a part of a fabric, researchers hope to destroy toxic bacterial agents when they appear. In addition, a catalytic system would destroy toxic chemical agents. When a soldier is exposed to an agent, fabrics made from nanomaterials could help to restrict the diffusion of the toxic agent by shrinking the size of the pores in the fabric. Wearing clothing that seals itself only when a toxic agent is present also allows moisture from perspiration to evaporate through the clothing most of the time. This means that soldiers won't feel like they're walking around in a sauna when toxic agents aren't present.

Researchers are also developing nanofiber mats that can filter a hazardous agent. The toxic chemical or bacterial molecules become attached to the nanofibers, while air is allowed to flow through the nanofiber mat. These mats could be used either as protective clothing or in face masks.

Helping sailors keep their propellers whirring

If propeller shafts on ships were investments, they'd be a bad one. Currently, the life of a propeller shaft is about only one year. No propeller coating can survive in a saltwater environment for more than a year because of galvanic corrosion, an electrochemical process that corrodes metal.

Four Navy ships are currently testing a new ceramic nanocomposite coating material that could withstand that corrosion. Made of nanostructured alumina titania, this coating has kept propeller shafts damage-free after five years of testing.

Creating lightweight portable power

As part of its effort to lighten a soldier's load, the U.S. Army is developing batteries using nanomaterial electrodes that increase the energy density (the amount of energy stored) in batteries. In Chapter 10 we discuss techniques of using nanomaterial-based electrodes in lithium-ion batteries to increase the surface area of the electrodes, and in some cases, change the chemical composition of the electrodes. These changes increase the amount of lithium ions that can attach to the electrodes, which increases the energy stored in the batteries. Estimates are that some of these techniques could result in batteries with five to ten times the energy density of standard lithium-ion batteries.

According to a presentation by researchers at the U.S. Army Research, Development and Engineering Command, they have two targets to hit:

- ✔ 200 watt hours per kilogram (which is about twice the energy density of standard lithium-ion batteries) in 2012
- ✔ 750 watt hours per kilogram around 2035

The Army is also planning to design the battery so that it can conform to a soldier's body. These batteries will be smaller, lighter, and longer-lasting than other batteries, as well as being flame retardant.

Making bulletproof material flexible

The term *liquid armor* refers to fabric in Kevlar vests that contains a liquid that, when struck by a bullet, becomes rigid. This kind of fabric contains a shear thickening fluid, or STF. STF is a suspension of hard nanoparticles in polyethylene glycol. This material is safe to wear against the skin and can be used in extreme temperatures.

Kevlar vests can be soaked in STF. The fabric then acts like any other fabric until it's hit by an object such as a bullet or shrapnel. According to Dr. Eric Wetzel, leader of a research team at The Weapons and Materials Research Directorate, part of the U.S. Army Research Lab: "During normal handling, the STF is very deformable and flows like a liquid. However, once a bullet or frag[ment] hits the vest, it transitions to a rigid material, which prevents the projectile from penetrating the soldier's body."

Although still under development, this liquid armor using nanoparticles holds the promise of better protection from bullets than current Kevlar fabric, while being thinner, more lightweight, and more flexible.

Part III
Nanotechnology and People

The 5th Wave   By Rich Tennant

THE NEXT EVOLUTION IN PALM-TOP ORGANIZERS

The PatchPilot

Delivers a preset amount of data into your bloodstream which quickly rushes it to your brain and subconscious.

I have to go now. I suddenly got the feeling I'm supposed to be at a meeting in 10 minutes.

In this part . . .

Beyond its technology aspects, nano brings up some interesting ethical and regulatory issues that those who enter the field must address in the future. This begs the question, just how do you break into this field to become part of those fascinating ethical discussions?

In this part, we explore the ethical, safety, and regulatory issues that may have an effect on how you interact with nanotechnology products or processes in your daily life. Then we look at the educational and career opportunities you might want to take advantage of to become part of this growing field.

Chapter 13

Nano Ethics, Safety, and Regulations

*N*anotechnology is often referred to as a *disruptive* technology because of its capability to produce drastic change. Disruptive technologies change the rules of the game because they open the door to drastically new ways of doing things that have a huge effect on our society. Examples of such a technology are the steam engine in the Industrial Age and the introduction of computers in the Information Age.

With disruptive technologies you get great rewards, as we discuss throughout this book. However, the drastic changes that technology can bring also have a downside. Three key areas of concern related to nanotechnology are

✔ Ethical questions raised by the advances that nanotechnology makes possible

✔ Safety concerns of using newly created materials

✔ Ways in which world governments could regulate the use of nanotechnology

In this chapter, we provide an overview of some of these concerns and several organizations keeping an eye on new developments. We also present some of the current thinking about nanotechnology in these three key areas.

Addressing Ethical Dilemmas

As mentioned, just about all the possibilities for using nanotechnology to improve our lives raise interesting ethical questions. If nano makes medical miracles possible, who should benefit from those miracles? If we could create anything we want for ourselves seemingly out of thin air, what would happen to our economy? And who is going to make choices about the priorities for how nano can help effect these changes?

Living forever

Nanotechnology could extend our lives in a couple of ways: by helping to eradicate life-threatening diseases such as cancer, heart disease, and diabetes and by making it possible to repair damage to our bodies at the cellular level — a nano version of the Fountain of Youth.

Perhaps the most exciting possibility for prolonging our lives exists in the potential for repairing our bodies at the cellular level. For example, as we age, DNA in our cells is damaged by radiation or chemicals. Nanotechnologists are developing techniques for building nanorobots that would be able to repair damaged DNA, allowing our cells to function optimally. The capability to repair DNA and other defective components in our cells goes beyond keeping us healthy: It has the potential to restore our bodies to a more youthful condition.

Cellular repair brings up several ethical questions. If nanotechnology helps us to live decades longer, what is the moral imperative for making such benefits available to all? If everybody could live hundreds of years, what would happen to our economy and society? Would only an elite few get such treatments, and what consequences would that have? If nobody ever died, would people have to stop having children to avoid overpopulation?

Several theories and opinions exist about the effect of life extension. A pro-life-extension article in the *Journal of Medical Ethics* (D. E. Cutas, March 2008) expresses one opinion about concerns related to overpopulation:

> *Whether or not the overpopulation threat is realistic, arguments from overpopulation cannot ethically demand halting the quest for, nor access to, life-extension. The reason for this is that we have a right to life, which entitles us not to have a meaningful life denied to us against our will and which does not allow discrimination solely on the grounds of age.*

The Center for Responsible Nanotechnology concurred with this approach for slightly different reasons. It published results of its research into how molecular manufacturing, one area of nanotechnology that offers options for cellular repair, might help extend our lives. They summarized one finding that supported moving forward with the molecular manufacturing (MM) area of nanotechnology in this way:

Overpopulation is a centuries-old problem. Traditionally, it's been solved by infanticide, plague, and vicious war. With MM, we'll have many decades to figure out better solutions. Smallpox vaccination and anesthesia were also said to be immoral. Today it's obvious that that's crazy. No one wants to be sick, and life extension is a natural result of health extension.

But not everybody is at ease with us living hundreds of years. At the 10th Annual Congress of the International Association of Biomedical Gerontology, this point was made about the desirability of generational turnover:

Is it preferable to have a group of people who live on and on, with very little population turnover, or to have a turnover of generations to bring in new ideas and new social developments — and if the latter, is that a reason to inhibit the development of life-extension? (attributed to John Harris)

Clearly this direction of nanotechnology research raises some interesting hopes as well as several difficult questions.

See Chapter 9 for more about medical advances involving nanotechnology and Chapter 5 for details on molecular manufacturing.

Producing goods from thin air

Other possibilities offered by molecular manufacturing also present some interesting challenges. Think about owning a replicator like the one on *Star Trek;* you could have anything you want at any time (say "Earl Grey tea," and it appears). Molecular manufacturing may make such technology possible by enabling us to build materials and products from the cellular level up. What would the molecular replicator, once developed, do to our society as we know it today? If people could simply produce many of the items they need themselves, what would motivate people to work for the things they want in life?

Molecular manufacturing could also have an effect on our global economy. It could spawn a dramatic shift that would completely change the way we do business, possibly putting billions of people out of work. Entire industries could become obsolete. At the same time, such advances could make it easy and cheap to produce powerful weapons.

But we don't have to wait for replicators to appear on the scene; molecular manufacturing could so streamline current manufacturing procedures that the production of materials such as weapons could become much more efficient in a few short years. Dr. K. Eric Drexler, a nanotechnology pioneer, is concerned about how the molecular manufacture of weapons could affect our world. He envisions tiny robots building products in desktop-sized factories long before replicators become a reality. According to Drexler:

> *A large-scale and convenient manufacturing capacity could be used to make incredibly powerful non-replicating weapons in unprecedented quantity. This could lead to an unstable arms race and a devastating war. Policy investigation into the effects of advanced nanotechnology should consider this as a primary concern, and runaway replication as a more distant issue.*

See Chapter 5 for more about advances being made in molecular manufacturing.

Making the right choices

Perhaps one of the biggest ethical challenges of nanotechnology and the opportunities it could bring is simply figuring out who will make the choices and how wise those choices will be. What should be the priorities of nano research? If third-world countries could be helped by improvements in energy production or water quality, should those basic needs come before the need of a middle-class person to replicate his own iPhone or to charge his laptop only once a month? Will the financial benefits of consumers willing to spend money on a product or service outweigh the needs of poor countries and starving children? And will these choices be made country by country or on a global scale?

The question of how advances in nano might benefit third-world countries is one of the most discussed nano issues today. A 2004 piece in *Science Daily* ("Nanotechnology's Miniature Answers to Developing World's Biggest Problems") is still highly relevant. The article, which addresses several areas of nanotechnology benefits, including healthcare and energy, states: "Most waves of technology can increase the gap between rich and poor, but the harnessing of nanotechnology represents a chance to close those gaps." The article goes on to note that several third-world countries, including India, Thailand, Mexico, and the Philippines, are working to develop their own nanotechnology initiatives to ensure that they are not left out in the cold.

All of which begs the question, is anybody orchestrating the potential of nano to help third-world countries? Folks at the Toronto Joint Centre for Bioethics, who note that nanotechnology could address many of the world's most critical problems, have stated that ". . . to our knowledge, there has been no systematic prioritization of nanotechnology targeted towards these challenges faced by the five billion people living in the developing world."

As to who will make the many ethical choices that nano presents, Jacob Heiler and Christine Peterson published a brief on the Foresight Institute web site that nicely sums up how we might ensure that nanotechnology benefits will be distributed fairly:

Ensuring that nanotech benefits humanity, rich and poor alike, is a matter of policy. Deliberate and early action can be taken by governments and non-governmental organizations to increase the odds that the benefits to nanotechnology are widespread, and don't needlessly exacerbate already large disparities.

Exploring organizations working on nanoethics

Greater minds than ours are hard at work trying to address all the ethical challenges of nanotechnology. Here is a sampling of those groups along with their URLs so you can visit their web sites to discover more about them:

✔ The Nanoethics Group (www.nanoethics.org) is an organization that studies the ethical and societal implications of nanotechnology. They encourage public dialog and work with nanotech groups to study ethical issues.

✔ The Center for Responsible Nanotechnology (www.crnano.org) is a nonprofit research group. They state that their focus includes "the major societal and environmental implications of advanced nanotechnology."

✔ The International Council on Nanotechnology (www.icon.rice.edu), whose web site is shown in Figure 13-1, focuses on global implications of nanotechnology. Their stated mission is "to develop and communicate information regarding potential environmental and health risks of nanotechnology, thereby fostering risk reduction while maximizing societal benefit."

Figure 13-1:
The
International
Council
on Nano-
technology
offers
articles,
news, and
resources.

✔ Latin American Nanotechnology and Society Network (www.estudios deldesarrollo.net/relans) is a group that works to create a forum for people to discuss nanotechnology development in Latin America. The group helps universities, governments, and other organizations communicate and examine "the political, economic, social, legal, ethical and environmental implications of nanotechnologies that are domestically developed, and/or in collaboration with foreign centers and institutions, and imported goods that contain nanocomponents."

✔ Various research groups such as the Whitesides Research Group (http://gmwgroup.harvard.edu/research_devecon.html) at Harvard University are looking at health issues in developing countries. The Whitesides group, whose web site is shown in Figure 13-2, is exploring nano and the disciplines of materials science, engineering, and biology and how they can address global issues with a focus on healthcare diagnostics and local energy needs.

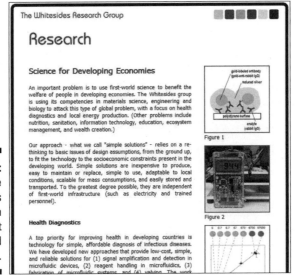

Figure 13-2:
The Whitesides Research Group at Harvard University.

Getting a Handle on Safety Issues

There's an old saying: Be careful what you wish for because you might get it. That advice could apply to nanotechnology. Could all the promise that nanotechnology holds be a double-edged sword, delivering miracles while causing unanticipated problems?

Looking at examples of safety concerns

Think of science fiction books you've read or movies you've seen. Some mad scientist applies a great new ray (or drug, or procedure) to a fly or a woman, and the fly turns into a man, or the woman grows to be the size of the Empire State Building. These scary scenarios exist in our science fiction literature for a reason: New science can bring surprising results. Because so much of nano-technology is new or still under development, various safety concerns have been raised, especially about the use of nanomaterials.

For example, when mice inhale carbon nanotubes, the material lodges in their lungs in a pattern similar to asbestos. What is not known is whether inhaled carbon nanotubes could cause cancer.

See Chapter 3 for more about carbon nanotubes and other nanomaterials.

If nanoparticles used in creams such as sunscreens could penetrate the outer layer of skin, would they cause damage to cells in the body? To find the answer, the U.S. National Center for Toxicological Research is conducting studies of the toxicity of the nanoparticles used in sunscreens.

Silver nanoparticles are useful in killing off bacteria and are already being used in household products such as cutting boards, detergents, and odor-controlling clothing. Currently, very few regulations about the uses of nanosilver exist. One study by researchers at Purdue University found that silver nanoparticles suspended in a solution were toxic to minnows. If silver nanoparticles released by detergents or other household products were to be released into our water supply, these incredibly tiny nanoparticles could not only kill fish but also move through fish egg membranes and kill off unborn generations.

Dr. Linda Birnbaum, the director of both the National Institute of Environmental Health Sciences and the National Toxicology Program, has made the following statement about the safety of nanomaterials:

> "We currently know very little about nanoscale materials' effect on human health and the environment. The same properties that make nanomaterials so potentially beneficial in drug delivery and product development are some of the same reasons we need to be cautious about their presence in the environment."

A statement in a paper by Na Gou, Analisa Onnis-Hayden, and April Z. Gu, from Northeastern University, is even more cautionary:

The recognized and unknown health risks and the harmful environmental impacts associated with the ever-increasing number of engineered nanomaterials in our daily life presents a serious threat to us all. This poses a pressing need for a breakthrough in toxicity-assessment technology because the available methods are neither feasible nor sufficient to provide the timely information needed for regulatory decision making to eliminate these threats.

Clearly there are serious and very real concerns about the safe implementation of nanotechnology and the use of nanomaterials in our world.

An overview of nanotechnology safety programs

To address some of these concerns, several organizations are setting themselves the task of watchdogging nanotechnology and related safety issues. Following is a list of some of those programs:

- **The Nanotechnology Core Facility,** created by the U.S. Food and Drug Administration's National Center for Toxicological Research, has a mission to "support nanotechnology toxicity studies, develop analytical tools to quantify nanomaterials in complex matrices, and develop procedures for characterizing namomaterials in FDA-regulated products." You can find more about this program by going to their web site (www.fda.gov/AboutFDA/CentersOffices/NCTR) and searching for *nanotechnology*.

- **The National Toxicology Program,** formed by the U.S. Department of Health and Human Services, has engaged in a research program whose purpose is described as follows: "to address potential human health hazards associated with the manufacture and use of nanoscale materials. This initiative is driven by the intense current and anticipated future research and development focus on nanotechnology." You can get more information by going to their web site (http://ntp.niehs.nih.gov) and searching for *nanotechnology safety initiative.*

- **The National Institute for Occupational Safety and Health** (www.cdc.gov/niosh/topics/nanotech/) has created its own field research team and directed it to "assess workplace processes, materials, and control technologies associated with nanotechnology and conduct on-site assessments of potential occupational exposure to a variety of nanomaterials."

- **The NanoHealth Enterprise Initiative** (www.niehs.nih.gov/research/supported/programs/nanohealth) was proposed by the National Institutes of Health to address critical research needs for the safe development of nanoscale materials and devices. According to the NIH web site, they propose "a partnership of NIH institutes, federal agencies, and public and private partners to pursue the very best science, leverage investment for research efficiencies, and minimize the time from discovery to application of engineered nanomaterials."

✔ **The Safenano Initiative** (www.safenano.org), whose web site is shown in Figure 13-3, was formed by the UK's Institute of Occupational Medicine. Their stated aims include a desire to "become the UK's premier independent site for information about Nanotechnology hazard, risk and good practice."

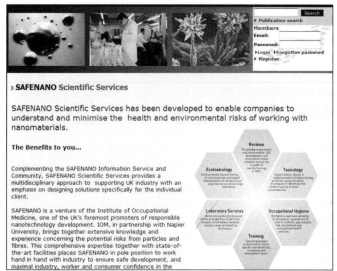

Figure 13-3:
The
Safenano
Initiative.

✔ **The Safety of Nano-materials Interdisciplinary Research Centre** (www.snirc.org) is a collaboration between the Institute of Occupational Medicine in Edinburgh, Napier University, Aberdeen University, Edinburgh University, and the U.S. National Institute for Occupational Safety and Health. Their goals include increasing awareness, providing information, and creating a body of scientific evidence that will help government and industry to create policies and procedures for the safe use of nanotechnology.

✔ **The Organization for Economic Co-Operation and Development** has a program to coordinate organizations in member countries for testing 14 key nanomaterials for human health and environmental safety. For more on this program, go to their web page at (www.oecd.org) and search for *testing manufactured nanomaterials.*

✔ **The Center for Environmental Implications of Nanotechnology** (http://cein.cnsi.ucla.edu), at UCLA, whose web site is shown in Figure 13-4, is focusing on how engineered nanomaterials could have an effect on cellular lifeforms in watery environments (both fresh and saltwater). Their stated mission is to be able to predict which nanomaterial physiochemical properties might be dangerous. After they identify these dangers, they then might be able to advise others on how to safely engineer nanomaterials to protect our environment.

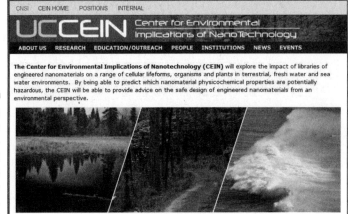

UCCEIN   Center for Environmental
Implications of NanoTechnology

ABOUT US   RESEARCH   EDUCATION/OUTREACH   PEOPLE   INSTITUTIONS   NEWS   EVENTS

The Center for Environmental Implications of Nanotechnology (CEIN) will explore the impact of libraries of engineered nanomaterials on a range of cellular lifeforms, organisms and plants in terrestrial, fresh water and sea water environments. By being able to predict which nanomaterial physicochemical properties are potentially hazardous, the CEIN will be able to provide advice on the safe design of engineered nanomaterials from an environmental perspective.

Figure 13-4:
The Center
for Environ-
mental
Implications
of Nano-
technology.

"Managing the Health and Safety Concerns Associated with Engineered Nanomaterials" is a report by the U.S. National Institute for Occupational Safety and Health that is useful for people who want to understand more about the safety and health issues that arise when nanotechnology is used in the workplace. This report, which you can find at `www.cdc.gov/niosh/docs/2009-125/`, includes a section on guidelines for working with engineered nanomaterials.

Scanning the Regulatory Landscape

As we emphasize in Chapter 1, nanotechnology isn't an industry, per se — it's a way to deal with materials on a tiny level. For that reason, nano goes across industries and often doesn't fit under any one regulatory body or set of rules. However, the need for regulation is real.

Understanding the need for regulation

The lead in your pencil is a larger scale or bulk version of carbon that has different properties than carbon nanotubes and buckyballs, even though all three share the same chemical composition. Because these bulk versions of materials are not considered hazardous, many regulatory agencies have not yet put in place different regulations for the use of nano-sized versions of the same substances.

However, because nanomaterials can have significantly different properties than the bulk form of the same material, different rules regarding the safe manufacture of everyday materials and their nano counterparts may be necessary. That's why governments are now evaluating special regulations for the use of nanomaterials.

Controlling the availability of nanotechnology could also be a challenge. In the case of molecular manufacturing and nano-sized factories (desktop-sized setups for manufacturing using nanomaterials), for example, the Center for Responsible Nanotechnology has stated

> *Uncontrolled availability of nanofactory technology can result from either insufficient or overzealous regulation. Inadequate regulation would make it easy to obtain and use an unrestricted nanofactory. Overzealous regulation would create a pent-up demand for nanoproducts, which if it gets strong enough, would fund espionage, cracking of restricted technology, or independent development, and eventually create a black market beyond the control of central authorities.*

The regulations governing nanomaterials will be evolving over the next several years. The following sections give a snapshot of this process in its early stages.

For updates on the latest nanotechnology regulatory changes, visit www. Understandingnano.com/nanotechnology-regulation.html.

Involving governments in regulation

Because nanotechnology is so promising, there is much government interest in regulating it. Here are a few highlights.

The U.S. Environmental Protection Agency (www.epa.gov/oppt/nano/) is developing a Significant New Use Rule (SNUR) which they state will

> *"Ensure that nanoscale materials receive appropriate regulatory review. The SNUR would require persons who intend to manufacture, import, or process new nanoscale materials based on chemical substances listed on the TSCA (Toxic Substances Control Act) Inventory to submit a Significant New Use Notice (SNUN) to EPA at least 90 days before commencing that activity. The SNUR would identify existing uses of nanoscale materials based on information submitted under the Agency's voluntary Nanoscale Materials Stewardship Program (NMSP) and other information."*

> *"The SNURs would provide the Agency with a basic set of information on nanoscale materials, such as chemical identification, material characterization, physical/chemical properties, commercial uses, production volume, exposure and fate data, and toxicity data. This information would help the Agency evaluate the intended uses of these nanoscale materials and to take action to prohibit or limit activities that may present an unreasonable risk to human health or the environment."*

Canada has put in place regulations to ensure that any new substance manu-factured in Canada or imported into Canada undergoes a risk assessment of its potential effects on the environment and human health. Environment Canadian has issued guidelines (www.ec.gc.ca/subsnouvelles-newsubs/default.asp?lang=En&n=3C32F773-1) to help determine whether a nano-material is considered a new substance.

The European Union is implementing a new Classification, Labeling, and Packaging (CLP) Regulation. CLP (http://ec.europa.eu/enterprise/sectors/chemicals/files/reach/nanos_in_reach_and_clp_en.pdf) includes the requirement that says that if the form or physical state of a sub-stance is changed by the use of nanotechnology, an evaluation must be con-ducted to determine whether the hazard classification should be changed. This could result in different classification and labeling requirements for bulk forms and nano forms of the same chemical substances.

The U.S. National Institute for Occupational Safety and Health (www.cdc.gov/niosh/topics/nanotech/safenano/) has published interim guidelines for working with nanomaterials in a report titled "Managing the Health and Safety Concerns Associated with Engineered Nanomaterials." NIOSH has created a Nanotechnology Field Research Effort to "assess workplace processes, materi-als, and control technologies associated with nanotechnology and conduct on-site assessments of potential occupational exposure to a variety of nanomaterials."

The U.S. Food and Drug Administration has the responsibility of reviewing many types of new products such as food additives and pharmaceuticals. They have summarized their stance on nanotechnology and such products on the web page titled FDA Regulation of Nanotechnology Products, which you can find by going to the FDA web page (www.fda.gov) and searching for *regulation of nanotechnology*.

Pulling in the private sector

Several nongovernmental organizations are also producing studies related to nano regulations, including the following:

 ✔ **The London School of Economics** has begun a Comparative Study of Nanotechnology project in which they are studying nanotechnology policies in some countries in Asia and the European Union.

✔ **The International Organization for Standardization (ISO)** has published a report called "Nanotechnologies — Methodology for the Classification and Categorization of Nanomaterials" that is intended to "promote clear and useful communication amongst industry consumers, governments and regulatory bodies." You can find the report on their web site (www.iso.org) by searching for *nanomaterials*. Be forewarned: This is only for those who are serious about their research because the report will cost you more than $100.

Securing the Promise of Nanotechnologies: Towards Transatlantic Regulatory Cooperation," is a report from the Royal Institute of International Affairs. This report summarizes the approaches toward regulation of nanomaterials in the European Union and the United States in an attempt to promote trans-Atlantic cooperation and consistency in the regulation of nanomaterials. You can view the report at www.chathamhouse.org.uk/files/14692_r0909_nanotechnologies.pdf.

Chapter 14

Making Nano Work for You: Education and Careers

In This Chapter

▶ Exploring the various areas of study for nanotechnology

▶ Discovering what degree options are available

▶ Understanding the nano job market

▶ Finding nanoworkers for your company

*I*f you're interested enough in nanotechnology to fork over the money for this book, you may just be interested in pursuing a career in nano. Your first step would be to get the education you need to follow your interest. To choose the right educational opportunity, you should also understand where such knowledge can lead you in the real-world job market.

In this chapter, we provide an overview of what's available in nanotechnology education and what career opportunities exist. And just in case your interest is in hiring employees, we also explore some of the ways that employers might connect with the nanotechnology workforce.

Be sure to also explore Chapter 16 for ideas about specific schools that have nanotechnology programs in place and what those programs offer.

Getting a Nano Education

Nanotechnology is something of an odd duck when it comes to choosing educational options because nanotechnology crosses over so many disciplines, from medicine to manufacturing. In addition, making the right choices for your nanotechnology education is tied into the career path you're considering. Do you want to be a nanotechnician doing work in a factory or lab, or do you want to become a researcher in a government, university, or corporate setting? The answer to that question will tell you whether you require an associate degree, a PhD, or something in between.

In this section, we explore several educational options to help you find your way.

To follow one student's experience with an intensive nanotechnology education experience at Penn State, check out this 30-minute video: www.pct.edu/degreesthatwork/nanotechnology.htm.

Which major is right for you?

Working in nanotechnology, with its focus on the characteristics of tiny pieces of matter, usually involves some knowledge of physics, chemistry, and biology. Any of these areas of study can provide a logical foundation for a nanotechnology education.

Some other areas of study that could have a nanotechnology angle include

- Agriculture
- Business
- Engineering (electronics, aerospace, mechanical, chemical, and so on)
- Environmental studies
- Ethics
- Forensic science
- Law
- Material science
- Medicine

In general, it's wise to choose a major that fits an area you'd like to work in, and get that basic education first . . . which brings us to choosing which degree to get.

What degree will get you where you want to go?

After you determine an area of study, you need to find a school that offers either a nanotechnology-specific degree such as nanomedicine or degrees in your area of interest, such as medicine or biology, with a focus or credential in nanotechnology.

Some schools offer degrees in nanotechnology, such as Penn State (one page of whose web site at `www.masters.nano.upenn.edu/` is shown in Figure 14-1); others provide a nanotechnology focus within degree programs such as chemistry, engineering, or medicine. You might consider getting an undergraduate degree in one of these areas, and then getting a graduate or postgraduate degree specifically in nanotechnology.

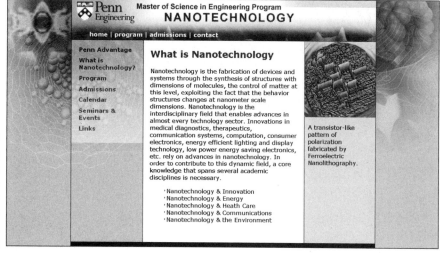

Figure 14-1:
Penn State's
Nano-
technology
Masters
program.

When you determine whether to go for a nano degree or simply a degree with a nano focus, how do you choose your degree(s)? With one projection of two million nanoworkers needed around the world by 2015, you can pretty much choose the level of education and career focus you want.

Start by deciding how much time and money you can invest in your career. According to NNIN, here are some common degrees with the associated time commitment for a full-time high school graduate:

- ✔ Technical program (2 years)
- ✔ Associate's degree (2 years)
- ✔ Bachelor's degree (4 years)
- ✔ Master's degree (6 years)
- ✔ Doctorate (8 years)

Technical school and associate's degrees will probably lead to nanotechnician jobs, performing somewhat routine chores in a fabrication or laboratory setting. Bachelor's degrees may lead a range of positions from process engineering in a manufacturing facility to marketing. Master's and doctorate degrees may lead to higher paying and perhaps more challenging jobs in research or academia.

Looking for schools with corporate alliances

Several schools are partnering with local companies to offer degrees or programs in nanotechnology that will meet those companies' workforce needs. In many such initiatives, companies are helping to create the curriculum, provide research facilities, and provide jobs for graduates.

One example is Dakota County Technical College (DCTC, at www.dctc.edu) in Minnesota. Working with more than 30 companies in their state, they developed the nanoscience technology curriculum shown in Figure 14-2. Because these companies helped to design the program, they are confident that graduates of the program will have the skills they need. Many of the companies have already recruited students from this program's ranks.

Dakota County Technical College is part of a program called NanoLink, which serves both educators and students of nanotechnology with information on partner colleges and opportunities for high school students interested in nanotechnology. You can visit the NanoLink program web site at www.nano-link.org/index.html.

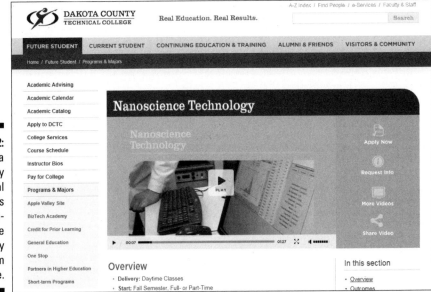

Figure 14-2: Dakota County Technical College's Nanoscience Technology Program web page.

In some cases, several schools have formed an alliance with corporations for their nanoeducation offerings. In Texas, the Nanoelectronics Workforce Development Initiative was developed by Austin Community College (ACC), Semiconductor Manufacturing Technology (SEMATECH, shown in Figure 14-3), and Texas State Technical College–Waco. Their focus is to train those interested in becoming engineers in nanoelectronics, a field vital to the semiconductor industry. This program includes the ACC NanoScholar Internship Program, in which selected students get work experience in Sematech's Advanced Technology Development facility (ATDF).

For more information about the Nanoelectronics Workforce Development Initiative and the ACC NanoScholar Internship Program, visit `www.sematech.org/research/nwdi/index.htm`.

A few other schools with corporate alliances are Rice University and the College of Nanoscale Science and Engineering (CNSE) of the University at Albany. We discuss both these schools in more detail in Chapter 16.

By finding a school with such corporate alliances, you could make your transition from nanoeducation to a paying job much smoother.

Figure 14-3: The Semi-conductor Manufacturing Technology web site.

Exploring government funded educational opportunities

Although the private sector is getting involved in nanotechnology education, both state governments and the U.S. Federal government see the need for a much more robust nanoworkforce in the next few years. According to Justine Johannes, senior manager in material science and engineering at Sandia National Labs, "We want to increase American competitiveness in nanotechnology. We want students to [pursue] degrees in engineering or science by exciting them with compelling problems and offering them the opportunity to make real progress toward solutions."

One example of a state-funded program is the ACC NanoScholar Internship Program, mentioned in the previous section. This program is funded by the State of Texas to give students a chance to get experience working in nanofabrication and nanotechnology research and development labs. Their model is to offer paid internships at the two-year technical, undergraduate, and graduate levels.

The U.S. Federal government is seriously invested in nanotechnology. As an important step in government involvement in nanotechnology education, in 2007 Congress passed the America Competes Act. This act spells out requirements for funding Innovation Institutes, which focus on science and engineering education with a goal of increasing the competitiveness of the United States in these areas.

These Innovation Institutes are associated with the National Institute for Nano-Engineering (NINE). NINE has a goal of putting nanotechnology in front of students to gain their involvement and to address three key themes: nano-electronics, nanoenergy, and nanomaterials manufacturing. You can see the NINE web site in Figure 14-4.

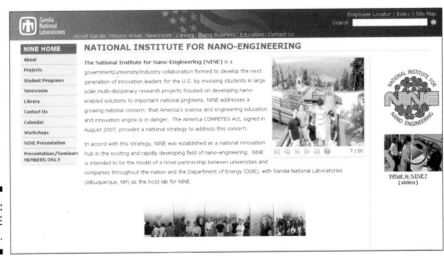

Figure 14-4:
The NINE
web site.

The entities associated with NINE reads like a private sector and educational who's who, including Corning Inc., Exxon Mobil Corp., Goodyear Tire and Rubber, IBM Corp., Intel Corp., Lockheed Martin Corp., Rensselaer Polytechnic Institute, Rice University, the University of California at Davis, Harvard University, MIT, Purdue, the University of Illinois, the University of New Mexico, the University of Notre Dame, the University of Texas at Austin, the University of Wisconsin, and Yale University. Sandia National Labs will coordinate the program, which you can learn more about at this URL: www. sandia.gov/NINE/home.html.

NINE was created in response to the National Academies report "Rising Above the Gathering Storm." This report states that the United States has to encourage technical education in college or risk falling behind developing nations in the number of scientists and engineers it produces.

Mapping Out a Career in Nano

The very good news is that the nanotechnology field has a phenomenal future. It spreads its fingers into almost every industry. It's so new that companies developing products and processes that use nano are scrambling to find qualified workers to help them succeed.

Jobs range from those who develop nanotechnology-based products to those who sell those products. In this section, we help you get a feel for the job potential and avenues you can follow to get yourself a career in nano.

Where's the need?

Both nanoscience and nanoengineering efforts will need a workforce to move them forward. According to Theodore von Karmam, who founded the Jet Propulsion Laboratory, "The scientist describes what is; the engineer creates what never was." Nanoscientists may perform research and postulate theories, whereas nanoengineers work with manipulating matter on the nanoscale. *Nanoengineering* is by nature interdisciplinary, sometimes involving chemistry, biology, or physics, for example, and often produces new products with never before seen properties and capabilities.

Many countries face a serious challenge in producing enough workers in the science and engineering area to support technologies such as nanotechnology. Some nanotechnology job projections estimate that nearly two million workers will be needed worldwide by 2015. The number of workers varies by country, but NNIN projections are

- 0.8–0.9 million in the United States
- 0.5–0.6 million in Japan
- 0.3–0.4 million in Europe
- 0.2 million in Asia Pacific (excluding Japan)
- 0.1 million in other regions

The numbers are actually bigger than those in the preceding list because nanotechnology will create an additional five million jobs in support fields and industries, according to one projection.

As these numbers reflect, the United States, in particular, has need of such people. Some studies have indicated that our economic growth is closely tied to technological advances; one study states that as much as 85 percent of our growth is due to technology. A weakening science and technology workforce will weaken our economy. As other countries are upping their contributions to the science and engineering labor pool, we have to work harder to catch up. Add to this the fact that our aging population is seeing the retirement of many in our scientific and engineering community, and the need to produce more workers in this area becomes urgent. A 2006 report in Nanotechnology Law and Business reported that "Now 29 percent of all science and engineering degree holders, and 44 percent of science and engineering doctorate holders, are greater than 50 years old. Over the next decade, the number of individuals with science and engineering degrees reaching traditional retirement age is expected to triple."

Richard Smalley himself predicted that eventually, 90 percent of all PhDs in the physical sciences would be Asian, and that half of those would be working in Asia.

Essentially these facts suggest that, though there is a need for nanoworkers around the world, the need may be greatest in the United States. Figure 14-5 shows a graph that compares U.S. scientific publications with the European Union and Asia Pacific countries, and indicates a decline only in the United States.

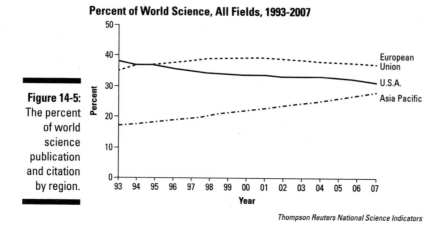

Percent of World Science, All Fields, 1993-2007

Thompson Reuters National Science Indicators

Figure 14-5:
The percent
of world
science
publication
and citation
by region.

Scoping out career opportunities

So, what industries will offer these nanocareer opportunities? Because nano-technology is part of so many processes and product types, a broad spectrum of industries need nano workers.

Current career opportunities fall into several areas where nanotechnology is being applied, including

- Aerospace industries
- Auto manufacturers
- Biotechnology
- Cosmetics
- Electronics/semiconductor industry
- Energy production
- Environmental monitoring and control
- Food science (both quality control and food packaging)
- Forensics
- Healthcare including diagnostics and treatments
- Lab research (government and academic)
- Materials science
- Military
- National security

- ✔ Pharmaceuticals, including drug delivery
- ✔ Retail (including RFID smart tags)
- ✔ Sports equipment

Our world will probably depend more and more on scientific advances to provide food and energy and to protect our environment in future years.

How much could you make?

According to Pennsylvania State University Center for Nanotechnology Education and Utilization, you can expect a rough range of salaries in nanotechnology jobs based on the degree you obtain:

- ✔ Two-year associate's degree: $30,000–$50,000
- ✔ Four-year bachelor's degree: $35,000–$65,000
- ✔ Six-year master's degree: $40,000–$80,000
- ✔ Eight-year doctorate: $75,000–$100,000

These salaries will vary depending on the industry and the job, but one thing is sure: These numbers will only increase in the future.

For more information about careers for nanotechnologists, visit www.nano. gov/html/edu/careers.htm. On this site, shown in Figure 14-6, you'll find links to relevant associations and career centers as well as some interesting links to articles on careers in nanotechnology. One word of warning: Because things change fast in nano and governments move slowly, sometimes links on this site won't work.

Exploring nano job services

In addition to more general job search sites such as www.jobrapido.com, some job services and online listings specialize in jobs with a nanotechnology spin. Here are a couple you might check out:

- ✔ Tiny Tech Jobs (www.tinytechjobs. com/). Although some of the content on this site is dated, they continue to offer current job postings and information on postgraduate fellowships in various nanotechnology-related fields. Employers listing job openings

pay a fee, and job seekers need to register with the site to respond to listings.

- ✔ Nanowerk (www.nanowerk.com/ nanocareer/homepage.php). This popular site provides free job postings that job seekers can search by keywords, such as job title, and sort by country, employer name, or most recently posted. If you are interested in a job posting, you can respond directly to the employer by clicking a link in a posting.

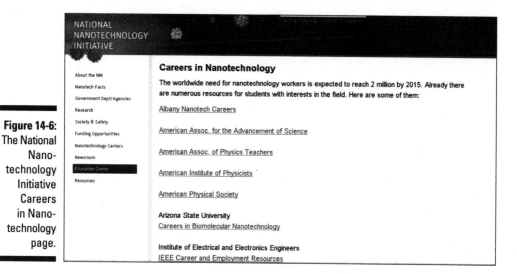

Figure 14-6:
The National
Nano-
technology
Initiative
Careers
in Nano-
technology
page.

Understanding what employers want

In polling the opinions of nanoemployers, we've found two distinct schools of thought. One group wants an interdisciplinary education resulting in nanotechnology credentials, but the other prefers people with a solid understanding of their specialty, such as chemistry or biology, with some nanotechnology courses included in their studies.

One reason for the latter approach may be that scientists and engineers in general learn the discipline of learning. If they get that skill, they can broaden their knowledge to related areas when needed to take on a new job or solve a perplexing problem.

Some companies are de facto in the business of nanotechnology, such as major manufacturers of computer chips. These products are all manufactured at the nano level. If such a company hires a promising engineer fresh out of school, for example, the company has all the tools necessary to help that engineer learn to use nanotechnology techniques for a particular position, whether that person has a degree in nanotechnology or not.

According to Justine Johannes of Sandia National Laboratories, "We want graduates to have more breadth and depth than they would likely have otherwise, so they learn how to work with partners on multidisciplinary teams as well as accumulate both technical and business experience."

Stan Williams, a nanotechnology researcher at Hewlett Packard, is often asked about how to prepare for working in the nanotechnology field. According to Williams, "I tell them to figure out what they like and get good at it, and to take communications courses, whether writing or journalism."

For our money, you should also consider communications courses such as public speaking. The ability to clearly and concisely communicate, whether written or verbally, with your coworkers, bosses, and customers can make a big difference in your career prospects.

Ken Smith of Carbon Nanotechnologies stresses the importance of interdisciplinary studies. According to Smith (specifically referencing Rice University's offerings), "Individuals with an educational background in these interdisciplinary areas are very few in number. Rice University's idea of combining a rare and highly demanded technical education with a modest exposure to training in business will produce students who are truly unique, and these students will be highly recruited by industry."

Desperately Seeking Nano Workers: Advice for Employers

Because nanotechnology is such a hot field, employers may have to scramble to fill their ranks with qualified people.

Just as those who want to work in nanotechnology might want to consider schools with corporate affiliations, your company might consider creating a partnership with a nearby college to ensure that you are on the spot when qualified grads come along. You might also think about setting up internships for local students and connecting with local high school classes to plant the seed for future workers.

Some employment services specialize in scientific workers, such as www.thesciencejob.com and www.physicstoday.org/jobs (shown in Figure 14-7), and some focus entirely on nano, such as www.tinytechjobs.com. Many online employment sites allow you to post job descriptions to try to reach potential employees either for free or for a fee.

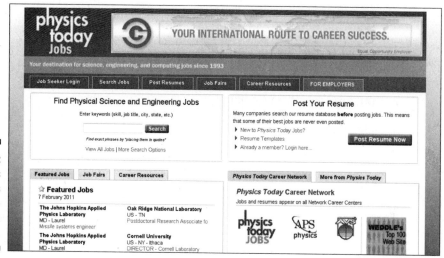

Figure 14-7:
Physics
Today's
employ-
ers' sign-in
page.

Consider attending nanotechnology conferences — such as the International Conference on Nanotechnology run by IEEE or Nanomaterials, a European conference focused on manufacturing and nanomaterials — to make connections.

But don't overlook those already in your employ. For example, engineers and scientists are trained to be able to learn new technologies and apply them. If you have somebody with engineering background performing one type of job, consider asking that person if he or she is interested in taking on a new job with a nanotechnology focus. Sometimes nanotechnology skills can be simply picked up on the job. Optionally, you might assign a mentor, send the employee to a conference or workshop, or even underwrite a few classes in nanotechnology at a local or online college.

Part IV
The Part of Tens

The 5th Wave   By Rich Tennant

"Not quite nano-size yet, but we're getting there."

In this part . . .

Because nanotechnology applies to many industries and settings, lots of people and organizations are involved in nano. From web sites like our own, which inform and educate people on nano, to universities offering degree programs and labs that provide invaluable resources to advance our knowledge of nano, a lot is going on.

In the three chapters in this part, we provide an overview of ten great web sites related to nano, ten universities that are offering interesting nano programs, and ten research labs at the forefront of nanotechnology research and development.

Chapter 15

Top Ten Nano Web Sites

In This Chapter

▶ Wandering around web sites that provide articles, news, and useful databases

▶ Having fun at a nano web site for the younger set

▶ Exploring web sites that represent government and healthcare issues

▶ Reviewing sites that stress nano safety and ethics

Some wonderful online resources are available to help you keep up on the latest nano advances, explore companies working with nano, discover what governments and research labs are up to, and more. Because nanotechnology changes just about daily, knowing which resources exist and what their strong suits are can save you time in finding the information you need.

In this chapter, we offer ten of what we consider to be the best web sites and call your attention to their coolest features to help you decide which ones might be of interest.

UnderstandingNano

www.understandingnano.com

Yes, UnderstandingNano is our site, so we're a bit prejudiced. However, the growing traffic to our site over the last few years, and the interest it has generated, especially among educators and students, makes us think it's worth your while to pay us a visit. Plus, we regularly post updated information to supplement this book and keep your nano knowledge current.

Here's what we consider the strengths of our site:

　✔ Perhaps the best collection of information on nanotechnology applications: We explore how nanotechnology is being used across a wide range of applications, including healthcare, manufacturing, energy, and the environment (see Figure 15-1).

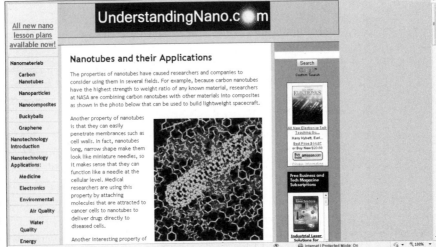

Figure 15-1:
Under-
standing-
Nano offers
information
on a wide
variety
of nano
applications.

✔ Nanotechnology lesson plans for middle school and high school students: These plans provide great tools for educators to help students learn about the basics of nanotechnology, as well as how nano is being used in medicine and to help our environment. Each lesson plan includes a student handout, reading assignments, discussion topics, and suggested projects.

✔ Directory of nanotechnology companies: We think this list is easy to use because we've divided it into categories such as Air Quality and Electronics to help you find the right company for your needs. Click a company to go to its site.

✔ A listing of nanotechnology degree programs: This list helps those considering a career in nanotechnology to find the right school for their studies.

Nanowerk

www.nanowerk.com

The interesting Nanowerk web site touts itself as the premier nanotechnology portal. The site is a collection of information about nanotechnology, ranging from news items to career listings and some informative introductory information for those new to the topic. Here are the things we like best about the site:

✔ The best nanomaterials database we've found: This is a useful, extensive, and easy-to-navigate system for finding providers of different kinds of nanomaterials. The results (see Figure 15-2) often give specifics about the size or purity of materials.

✔ Excellent "Spotlight" articles: These articles provide a detailed review of recent nanotechnology developments. Most articles are written in collaboration with authors of scientific papers, so the content is usually well grounded in scientific know-how.

✔ A thorough collection of nanotechnology news stories: This site tends to include just about every piece of news on nanotechnology, so you may have to troll through to find the most current and interesting items.

✔ Nanotechnology degree program directory: The programs are divided into Bachelor/Master/PhD/Other, and you can view the results by country or alphabetically.

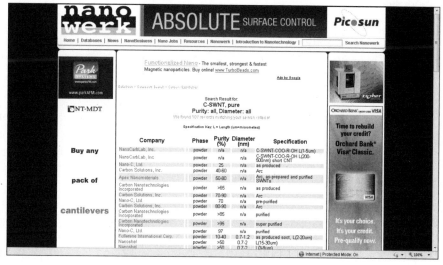

Figure 15-2: Nanowerk. com's nano-materials database can be a great resource.

NanoZone

www.nanozone.org

The NanoZone site is for the kid in all of us. It provides well-prepared introductory materials on nanotechnology for students in the 8- to 14-year-old range. In addition, some sections provide information for museum professionals and teachers.

Here's what we especially like about Nanozone:

- ✔ Short podcasts highlight what is special about nanotechnology (click Why, and then click Talk to a Scientist to find them).

- ✔ Videos (for example, under How Small Is It?, What Is a SEM?) allow you to see things at the nanoscale and hear how people are working with nanosized materials. You can also find images sprinkled around the site of various materials viewed using a scanning electron microscope.

- ✔ Ads of the future (under Why Is It Important, click FUTURENANO) are not only fun to look at but also tell you when certain products might be available and what makes them nanoproducts. This feature is shown in Figure 15-3.

- ✔ NanoStat cards for various scientists (click Who Works On It) show you the background of various people who became nanotechnologists. This section provides a good way for those interested in nanotechnology to understand the kind of studies and pursuits that might lead them to success in the nanotechnology field.

The big drawback to the site is that it was last updated in 2006 and seems to be a project that has ended, so beware any dated material. But as a fun site on nano basics, it should be useful to many.

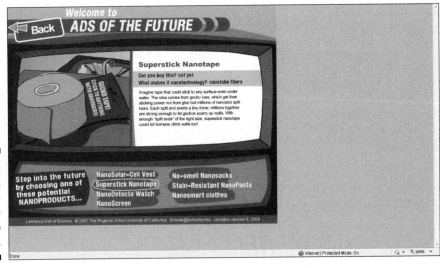

Figure 15-3:
Nanozone's ads for pos-sible future nano products.

National Cancer Institute Alliance for Nanotechnology in Cancer

www.nano.cancer.gov

The National Cancer Institute Alliance for Nanotechnology in Cancer is a prime example of a government funding university-based research centers. These centers develop techniques that result in startup companies working to commercialize products. Some of those results include an anticancer drug that causes cancer cells to kill themselves off and unique ways of using nano for imaging to detect some forms of cancer. To see a longer list of examples of these projects, look under Alliance in Action and view their Accomplishments.

Watching their video titled "Journey into Nanotechnology" is one way to understand how confident this alliance is that their work will help prevent cancer. They also offer some introductory material to help you better understand what nanotechnology is; it features some very nice illustrations of things such as nanoshells and nanowires (see Figure 15-4). Their Tools for Education section is also a great resource for videos and images to help people of all ages learn about nano and its use in detecting or curing cancer.

You can also find news about the latest advances in nanotechnology cancer treatments on their News and Highlights page.

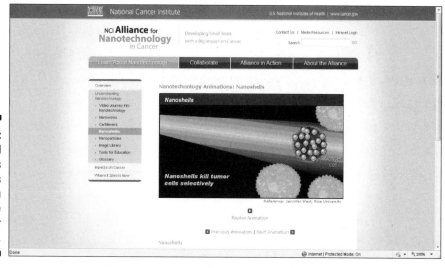

Figure 15-4:
Colorful illustrations and videos help you visualize nano cancer progress.

Foresight Institute

www.foresight.org

Foresight Institute is a self-professed "think tank and public interest organization" that offers a series of roadmaps to move forward in molecular manufacturing as well as educational content and advocacy for the ethical use of nanotechnology products and techniques.

Here are some of the highlights of this site:

- ✔ In one of the most extensive nanomedicine image galleries available, you can find images of a wide range of nanorobots and other medical-related nanomachines (see Figure 15-5).

- ✔ The Why Should You Care? link takes you to an interesting group of essays by scientists, including nanotechnology luminaries Drexler and Freitas, who comment on why you should care about nanotechnology.

- ✔ A group works with others to attempt to standardize terminology, forms of measurement, health and safety standards, and the way that materials are specified.

- ✔ The Open Source Sensing Initiative is an intriguing attempt to balance the need to use nanotechnology-based sensors to collect health and other data with the need to respect individual privacy.

Figure 15-5:
The Foresight medical imaging gallery is one of the best out there.

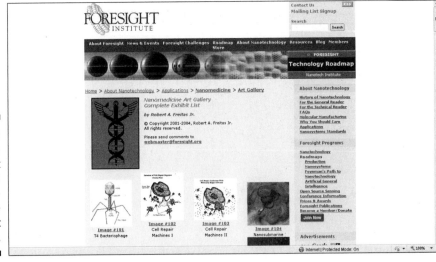

Nanoforum

http://nanoforum.org/

Nanoforum calls itself the European nanotechnology gateway. The site boasts a few dozen well-written Nanoforum Reports, which come out every few years. Some reports have a European slant, such as "Nano-education from a European Perspective" and "Nanotechnology in the Nordic Region." Others, such as "Nanotechnology in Aerospace" or "Intellectual Property in the Nanotechnology Economy," take a broader perspective but still with a European slant.

You can also find several useful abstracts of recent nanorelated publications compiled from various European resources such as the Centre for Science, Society, and Citizenship in Rome, Italy.

Finally, a rather fun introduction to nano basics for children comes in the form of the Nano Education Tree. By exploring the branches of this tree (see Figure 15-6), which is on the What Is Nano? page, younger folks can learn about nanotechnology in energy, electronics, health care, and more.

Figure 15-6:
Nanoforum's Education Tree offers a pleasant graphical way for students to explore nano topics.

National Nanotechnology Initiative

`www.nano.gov`

National Nanotechnology Initiative, a 10-year-old government initiative, coordinates federal agencies that fund a group of prestigious research centers located around the United States. The site provides information about the centers and their work, and includes a strategic plan that is updated every few years. The plan outlines their vision and the current priorities and goals.

You'll also find a centralized listing of Nanotechnology Research Centers and Networks funded by the U.S. government, as well as a useful nanotechnology degree program list for those interested in nano as a career.

The Society & Safety section of the site offers government perspectives on occupational safety, environmental issues related to nanotechnology, and the potential effect of nano on our society.

See the Resources section for links to the text of Nobel Prize winners' lectures on nanotechnology (see Figure 15-7). The lecture pages also offer biographical information about the winners, photo galleries, and, in some cases, interviews.

Figure 15-7:
Learn from
the winners
on www.
nano.gov.

NanoTechnology Group

www.tntg.org

The NanoTechnology Group site is the best resource we've found to help teachers find nanotechnology education activities that may be useful in their classrooms.

Here are some of their best offerings:

- Their listing of sources of virtual and simulation tools is useful for helping students visualize nanotechnology tools and processes.
- Games for Education is a great resource for links to educational but entertaining tools such as NanoMission or the Noble Prize educational games site (see Figure 15-8).
- The K-12 Education Outreach section of the site lists resources, such as our own lesson plans on UnderstandingNano.com, that educators can use in developing lessons or curriculum to teach students about nanotechnology.

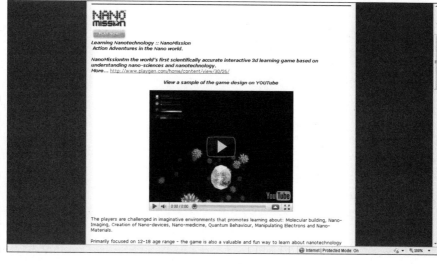

Figure 15-8: Links on the Nano-technology Group site lead to lots of interesting educational resources.

Safenano

www.safenano.org

Safenano, a UK-based site, offers the best collection of news about developments in nanotechnology safety we've found online. You can join discussions about safety or ask a question. They also have a database of back articles with a search engine that helps you narrow your search to get the articles most relevant to your needs.

Here are some other features we like on this site:

- ✔ The Current Awareness section of the site provides articles, press releases, and listings of conferences related to nanotechnology and safety issues.

- ✔ SafeNano Scientific Services provides advice to businesses facing safety challenges related to nano. This service is a joint effort between the Institute of Occupational Medicine and Napier University.

Note that you do have to register to use most of the site features, but registration is free.

Nanotech-now

www.nanotech-now.com

The Nanotech-now site hosts one of the best nanotechnology image galleries of images that address a wide range of nano topics (see Figure 15-9). This gallery includes photos of nanoscale objects and illustrations of proposed nanorobots and art pieces in which an artist has taken a photo of a nanoscale object and manipulated it to produce a piece of abstract art.

Cristian Orfescu is one pioneer in this field of art and has even organized nanoart competitions. See his work at www.crisorfescu.com. Another artist, Nicolle Rager Fuller, produces realistic illustrations of nanotechnology at work. You can see her images by visiting www.sayo-art.com/portfolio/small.php#.

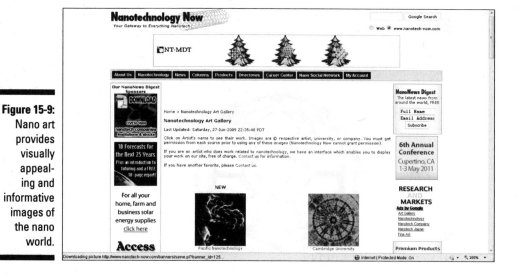

Figure 15-9:
Nano art
provides
visually
appeal-
ing and
informative
images of
the nano
world.

Nanotech-now is also one of the best sites for finding nanotechnology news stories and press releases. Click their News tab for stories, press releases, and interviews with various folks in the nanotechnology field. The Press Kit offered in this section is a useful resource for those writing about nanotechnology, providing information on the history, basic facts, and risks and benefits of nanotechnology, as well as a useful glossary.

The columns on Nanotech-now are written by experts from a wide range of nanotechnology disciplines from nanoinvesting to nanoethics and nanosolar.

Chapter 16

Ten Nano Universities

In This Chapter

▶ Exploring graduate and postdoctoral education at world-class educational institutions

▶ Taking a look at opportunities for AA and BA degrees

▶ Discovering how schools are offering students access to facilities and corporations

*N*anotechnology education is in its relatively early days because the field itself is so new. However, you can find lots of interest and activity out there. From AA and BA degrees that make sense for those who want to pursue careers as nanotechnicians or nanoproduct salespeople, to Masters and PhD programs for future researchers or academicians, nanoeducation runs the gamut.

In this chapter, we cover ten schools, ranging from a community college to Rice University, where the buckyball was first discovered, and from U.S. schools to some foreign universities. These are not all necessarily the best nor the biggest schools, but they should give any reader interested in getting an education in nanotechnology a nice variety of choices. All the schools included here have a strong focus in nanotechnology programs and, in most cases, great nanotechnology resources in the form of labs or corporate affiliations.

The College of Nanoscale Science and Engineering (CNSE) at the University of Albany

www.cnse.albany.edu

We choose to highlight CNSE at the University of Albany because it offers three levels of degrees and has strong corporate affiliations with more than 250 companies, as well as a great research lab. CSNE includes centers with focuses in areas such as

✔ Energy

✔ Environment

✔ Lithography

✔ Nanoelectronics

✔ Nanomaterials

✔ Semiconductors

CNSE, whose web site is shown in Figure 16-1, has a strong focus on innovation and commercialization, meaning that they work with corporate partners to come up with ideas, research those ideas, and turn the resulting technology into commercial products. They are involved with the funding of product development and with licensing discoveries made by their lab. For those interested in this so-called real-world approach to scientific discovery, this school has much to offer.

Figure 16-1:
The CNSE
web site.

Follow these links for more about their degree programs:

✔ Bachelor's program: www.cnse.albany.edu/
PioneeringEducation/UndergraduatePrograms.aspx

✔ Master's and PhD: www.cnse.albany.edu/PioneeringEducation/
GraduatePrograms.aspx

For more information about their NanoTech Complex, go to
this link: www.cnse.albany.edu/WorldClassResources/
CNSEAlbanyNanoTechComplex.aspx.

Rice University

```
http://cnst.rice.edu/nano_at_rice/
```

As the home of the buckyball and longtime employer of two of its discoverers, Richard Smalley and Robert Curl, Rice has traditionally been a strong force in nanotechnology. With Smalley's passing in 2005, the Smalley Institute (`http://cnst.rice.edu`), whose web site is shown in Figure 16-2, carries on his work. This institute has 151 faculty and staff members who represent 21 departments. More than 600 students study nanotechnology through its auspices.

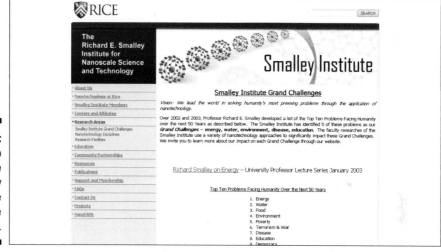

Figure 16-2: The web site of the Smalley Institute at Rice University.

Rice's Center for Biological and Environmental Nanotechnology (CBEN), whose web site is `www.cben.rice.edu/`, focuses on nanotechnology in medicine and the environment. The Center has stated that part of its focus is to draw new talent into the field of nanotechnology. As part of that focus, they work with teachers in the Houston School District to train them in what they call a "discovery-based teaching style." They also work with students at Rice at both the undergraduate and graduate levels through their Center Science Academy. Finally, they have a partnership with the Jones Graduate School of Management to address the needs of small startup technology companies.

One great example of Rice's affiliation with industry is the Lockheed Martin Advanced Nanotechnology Center of Excellence at Rice University (`www.lancer.rice.edu/`). Rice describes this association as "a unique nanotechnology research program to explore new technologies for materials, electronics, energy, security and defense. Through LANCER, Lockheed Martin engineers will pair with Rice experts in carbon nanotechnology, photonics, plasmonics and other nanoscience disciplines to address a broad range of potential nanotechnology applications."

Rice offers an MS in Nanoscale Physics. You can get more information about this degree program at www.profms.rice.edu/nanophysics.aspx?id=64.

The university doesn't offer a nanotechnology PhD degree per se. Rather, students choose a field of graduate studies, such as natural science or engineering, and then work with a professor who has expertise in nanotechnology.

Northwestern University

www.northwestern.edu

They say where there's smoke, there's fire, and perhaps that's why Northwestern University is so hot. It seems like everywhere you go in the world of nanotechnology, this school is talked about. The only nano-specific degree they give is an MS, which you can learn more about at this link: www.mech.northwestern.edu/web/docs_info/documents/ms_flyers/NU_MS-Nanotech.pdf.

Where Northwestern really shines is in its resources for students: the International Institute for Nanotechnology and three important nano centers. Here's a bit more detail about each.

The International Institute for Nanotechnology (www.iinano.org/) started in 2000 and coordinates more than $500 million in nano research funding and infrastructure. IIN has partnerships in 18 countries outside the United States, as well as connections with universities worldwide. For students, these relationships mean they can take advantage of postdoctoral exchange programs. Their annual symposium draws researchers from around the world.

IIN has helped to launch 16 startup companies through their Small Business Evaluation and Entrepreneur program (SBEE). Their Nanotechnology Corporate Partners program consists of 20 member companies.

Established by funding from the National Science Foundation, the Nanoscale Science and Engineering Center (NSEC) for Integrated Nanopatterning and Detection Technologies (www.nsec.northwestern.edu/) has a vision of developing "innovative biological and chemical detection systems capable of revolutionizing a variety of fields." Detection techniques for Alzheimer's and prostate cancer, for example, have been advanced by the work of NSEC. The Center allows undergraduate and graduate students to take advantage of their research facilities. The Center also promotes education in nanotechnology through several interesting programs for teachers and precollege classrooms.

NSEC's exhibit on nanotechnology with the Museum of Science and Industry in Chicago is scheduled to open in 2011 and should provide some groundbreaking interactive exhibits for those who visit it.

Northwestern University Center of Cancer Nanotechnology Excellence (`www.ccne.northwestern.edu/`) is a wonderful resource for students interested in nanomedicine, specifically detection and treatment of cancer. This center is funded by the National Cancer Institute and works closely with the International Institute of Nanotechnology. CCNE offers several valuable research facilities on Northwestern's campus.

The Center for Quantum Devices (`http://cqd.eecs.northwestern.edu/`) was founded in 1991 with a focus on semiconductor research that involves nanotechnology. The Center works with the help of approximately 15 undergraduate and graduate students, as well as offering mentored research time to undergraduates at Northwestern. Their web site, shown in Figure 16-3, offers some interesting information and images of nanotechnology.

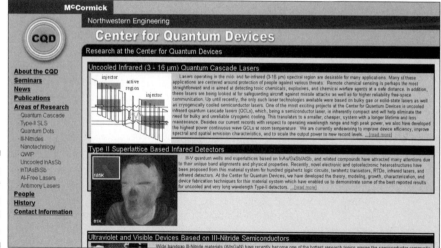

Figure 16-3:
The Center
for Quantum
Devices
web site.

Northeastern University

`www.northeastern.edu`

For those interested in using nanotechnology in the healthcare field, Northeastern offers a PhD in Nanomedicine. The program is made possible through an initiative called IGERT (the National Science Foundation's Integrative Graduate Education and Research Traineeship) with support from the National Cancer Institute. The program boasts a "tightly-integrated interdisciplinary team of medical researchers, pharmaceutical scientists, physicists, chemists, and chemical engineers" who help students with nano-medicine research and studies. The school stresses the interdisciplinary nature of studies, as well as opportunities for real-world experience and international research options. You can learn more about the IGERT program at `http://www.igert.neu.edu/research.htm`.

Northeastern offers two other resources for students interested in nanomedicine:

- ✔ The Center for Pharmaceutical Biotechnology and Nanomedicine (`www.northeastern.edu/pharmsci/research/centers/center/`) grants MS and PhD degrees in Pharmaceutical Science. They offer a focus on Drug Delivery/Nanomedicine, as well as other areas of focus with logical ties to nanotechnology, such as imaging.

- ✔ The Center for Translational Cancer Nanomedicine was approved in late 2010, so it won't be active for a few years, but its plans include finding a path from research to actual nanomedicine cancer cures available for clinical use. The Center is funded by a grant from NCI and will be led by the current director of the Center for Pharmaceutical Biotechnology and Nanomedicine.

Finally, shown in Figure 16-4 is the Nanoscale Science and Engineering Center for High-rate Nanomanufacturing (`www.northeastern.edu/chn/`), a great resource for students interested in nanomanufacturing, including large-scale directed assembly and transfer, environmental health and safety, and regulatory and ethical issues.

Figure 16-4: The Nanoscale Science and Engineering Center web site.

Dakota County Technical College

www.dctc.mnscu.edu/future-students/programs/nanoscience-technology.cfm

Not everybody who is interested in nanotechnology wants to be a Nobel Prize winner. Some of us just want to get our foot in the door in this exciting industry. The Dakota County Technical College caught our eye because they offer an associate's degree in Nanoscience Technology and outreach to support nanoscience classes in high schools. According to their web site, shown in Figure 16-5, "This program prepares students for careers in the nanobiotech, nanomaterials and nanoelectronics industries. Offered through a partnership with the University of Minnesota, the program gives graduates the skills and knowledge to land jobs in companies and corporations applying nanotechnology to product development, testing, research and development, and manufacturing design."

DCTC also has more than 25 partners in the corporate sector that offer both lecturers and an internship program. Through an association with the University of Minnesota, students may be able to access lectures and laboratory time at that school's Nanofabrication Center.

Figure 16-5:
The Dakota County Technical College web site.

As evidence of the success of this technical college's role in nanotechnology education, the National Science Foundation awarded them a $3 million grant to develop the Midwest Regional Center for Nanotechnology Education, or Nano-Link.

University of Waterloo

`www.uwaterloo.ca/`

With programs in nanoelectronics and nanoengineering, the University of Waterloo in Waterloo, Ontario, Canada, runs the largest nanotechnology degree program in Canada, offering BA, MA, and PhD degrees. The school was the first in Canada to offer an undergraduate nanotechnology engineering program, as well as a Masters in Business, Entrepreneurship, and Technology. Their involvement as host to the Canadian Water Network, a water quality research program, and their status as the first school in Canada to hire environmental faculty sets them up to provide students with opportunities to explore the use of nanotechnology to solve environmental problems.

The school has two research labs that help support student research and study:

✔ The Advanced Technology Laboratory (`www.nano.uwaterloo.ca/facilities/watlab.html`) whose web site is shown in Figure 16-6, is involved in research in areas such as nanofabrication, microscopy, and nanolithography. They offer research facilities to university and corporate researchers. Graduate students can both qualify to perform research and apply to get a WIN fellowship to help pay for their education.

Figure 16-6:
The
Advanced
Technology
Laboratory
at the
University of
Waterloo.

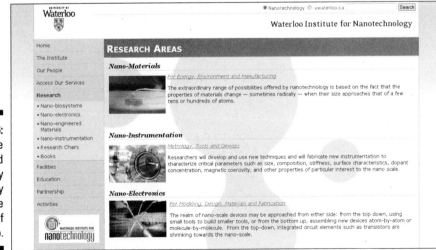

✔ The Nano-and Micro-systems Research Lab (`http://biomems.uwaterloo.ca/research.html`) is involved in research in nanoelectromechanical systems (NEMS), as well as micro assembly and the use of nanodevices for biomedical applications. The lab actively solicits undergraduate and graduate student researchers to work on a variety of projects.

To get more information about their degree programs, visit these web sites:

✔ Bachelor of Applied Science (BASc) degree in Nanotechnology Engineering: `www.nanotech.uwaterloo.ca/Undergraduate_Studies/`

✔ MASc, MSc and PhD programs in Nanotechnology: `www.nano.uwaterloo.ca/education/grad.html`

✔ Giga-to-Nanoelectronics (G2N) Laboratory: `www.nano.uwaterloo.ca/facilities/giga_nano_lab.html`

Universities of Leeds and Sheffield

`www.nanofolio.org/`

Nanofolio is a joint program between the Universities of Leeds and Sheffield. These schools tout that they have "the longest established nanotechnology masters courses in Europe." Their courses fit the bill for students from most scientific and engineering programs and fall into four degree areas:

✔ Masters (MSc) course in nanoscale science and technology (`www.nanofolio.org/courses/nst.php`)

✔ Masters (MSc) course in nanoelectronics and nanomechanics (`www.nanofolio.org/courses/nem.php`)

✔ Masters (MSc) course in nanomaterials for nanoengineering (`www.nanofolio.org/courses/eng.php`)

✔ Masters (MSc) course in bionanotechnology (`www.nanofolio.org/courses/bio.php`)

The resources they offer students include several recognized centers and labs for research, including the following:

✔ Centre for Molecular Nanoscience (`www.cmns.leeds.ac.uk/`): This Centre is located at the University of Leeds and has a focus on bottom-up nanotechnology including molecular self-assembly. This Centre offers opportunities for postgraduate students to work with a strong academic team on their research.

✔ **Centre for Nano-Device Modeling** (`www.maths.leeds.ac.uk/cndm/`): This Centre is also located at the University of Leeds and coordinates nanotechnology device modeling across the departments of applied mathematics, electronic and electrical engineering, physics, and astronomy.

✔ **Sheffield NanoLab** (`www.nanolab.org.uk/`): This lab provides research facilities for nanorobotics, nanomanipulation, nanometrology, nanotomography, nanoparticles, and nanomaterials characterization. PhD students at the University of Sheffield can take advantage of the lab's resources.

Nanometrology has to do with measuring things at the nanoscale. Nanotomography involves using x-rays to examine cross sections of materials to create virtual 3-D models, where the cross sections are taken at the nanoscale.

✔ **Leeds EPSRC Nanoscience and Nanotechnology Facility** (`www.nanotechnology.leeds.ac.uk/lennf/`): This facility at the University of Leeds offers a wide range of nanocharacterization and nanofabrication equipment. Access to their facility is limited to researchers who have or are eligible for an EPSRC (Engineering and Physical Sciences Research Council) grant. Their web site is shown in Figure 16-7.

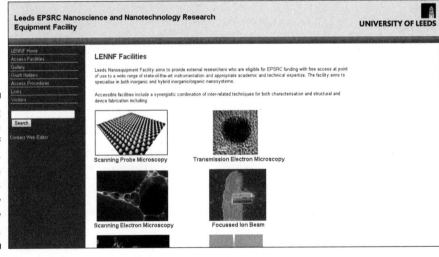

Figure 16-7: The University of Leeds Nanoscience and Nanotechnology Facility web site.

University of Washington

www.washington.edu

If you can overlook the rain, you might consider Seattle for your nanoeducation by looking into the University of Washington. The Center for Nanotechnology at UW (www.nano.washington.edu/about/index.html) offers a PhD in Nanotechnology, and gathers more than 75 professors from a variety of disciplines, such as chemistry, physics, bioengineering, electrical engineering, genome sciences, and microbiology. They are funded by the National Science Foundation's Integrative Graduate Education and Research Traineeship (NSF-IGERT) program.

UW has several resources in nanotechnology of which students can take advantage, including the following:

✔ The Nanotech User Facility (https://depts.washington.edu/ntuf/) gives Pacific Northwest researchers and companies an opportunity to use sophisticated characterization and nanofabrication tools. In 2004 the facility became one of the 13 nodes of the National Science Foundation's *National Nanotechnology Infrastructure Network*. Their emphasis is on nanotechnology in biology and life sciences.

The Nanotech User Facility at UW coordinates with the NSF Ocean Observatories Initiative to lead the way in exploring underwater sensor activity made possible through recent developments in nanotechnology.

✔ The Nanotechnology Modeling Lab (http://dunham.ee.washington.edu/) is a part of the Electrical Engineering Department at the College of Engineering. They have a focus on nanofabrication, with efforts centered on modeling, simulations, and fabrication of devices. Both graduate and undergraduate students have the possibility of working and researching in the lab.

✔ The University of Washington and Pacific Northwest National Lab (PNNL at http://www.pnl.gov/), whose web site is shown in Figure 16-8, established the Joint Institute for Nanoscience on the university campus in Seattle in 2001. A joint board approves awards of funding to promote joint projects. Awards are given for graduate student, postdoc, or faculty salaries; tuitions for grad students; and travel or housing expenses. The institute also offers courses for undergraduate and graduate students at the Environmental Molecular Sciences Laboratory (EMSL, at www.emsl.pnl.gov). EMSL is a Department of Energy national scientific user facility at Pacific Northwest National Laboratory in Richland, Washington.

Figure 16-8:
Pacific
Northwest
National Lab
web site.

Joint School of Nanoscience and Nanonengineering

```
http://jsnn.ncat.uncg.edu/
```

The Joint School of Nanoscience and Nanoengineering caught our attention because it's part of a local economic stimulus program with the strong involvement of startup companies. One of the two schools involved, North Carolina A&T State University, began as a school for African Americans, while the other, the University of North Carolina at Greensboro, was historically a women's school.

This joint school offers a Master of Science in Nanoscience (`http://jsnn.ncat.uncg.edu/academic/nanoscience/pms-nanoscience.html`), as well as a PhD in Nanoscience (`http://jsnn.ncat.uncg.edu/academic/nanoscience/phd-nanoscience.html`).

Based in North Carolina, JSNN offers courses in nanotechnology that were previously hard to come by in that area of the country. The school has begun teaching classes while building a facility that will be completed at the end of 2011.

According to the school description, it "seeks to develop collaborations with the local and regional businesses that will raise the Triad's Nanotechnology profile with the goal of attracting new industry and investment to the area and by doing so helping to stimulate the economic growth."

The folks at Pennsylvania State University and others are working to get nano-technology on the agenda for all high schools and community colleges throughout Pennsylvania. This effort is part of a workforce development program to support high-tech companies in Pennsylvania. To learn more about this interesting effort, called the Pennsylvania Nanofabrication Manufacturing Technology Partnership, visit www.cneu.psu.edu/edFAQs.html.

Stanford University

If Northern California and a world-class school are your cup of tea, check out Stanford. This school offers some great resources. Although they don't offer degrees specifically in nano, they do offer nanotechnology research opportunities for students who are obtaining degrees in other areas such as engineering or physics. Students produce a thesis or dissertation in a nano aspect of that field.

According to Jim Plummer, Dean of the School of Engineering, "We want to give our students and faculty ample opportunity to play in this new sandbox, where matter is manipulated at atomic and molecular scales. We will judge the Nanoscience and Nanotechnology priority successful if we find our students and faculty across a wide range of disciplines to be frontrunners in their respective fields."

Stanford's resources include the following:

✔ Stanford Nanocharacterization Lab (www.stanford.edu/group/snl/) is a National Nanotechnology Infrastructure Network facility. Its equipment and tools are available to all in the Stanford community who meet their qualifications, as well as to organizations that partner with Stanford. The focus here is on characterization of nanomaterials.

✔ Stanford Nanofabrication Facility (http://snf.stanford.edu/) is a great resource for nanoscience and engineering students and researchers to work with advanced nanotechnology fabrication and process tools. They offer an undergraduate NINN Research Experience program each summer that gives students a feel for what the graduate school offers.

✔ Center for Magnetic Nanotechnology (www.stanford.edu/group/nanomag_center/), shown in Figure 16-9, states that its mission is "to stimulate research at Stanford in the area of magnetic nanotechnology, magnetic sensing, and information storage materials, to facilitate collaboration between Stanford scientists and their industrial colleagues, to train well-rounded and highly skilled graduate students, and to develop curricular offerings in the relevant subjects." They have interesting offerings in the form of workshops, short courses, and conferences on nanomagnetics.

✔ Center for Probing the Nanoscale (http://www.stanford.edu/
group/cpn/) was created in partnership with IBM with a grant from
the National Science Foundation. As the name suggests, their area of
research is to develop probes that can help us see and manipulate
things at the nano level. They provide lots of support to teachers of
nanotechnology, including their Summer Institute for Middle School
Teachers. If you have an interest in teaching nano, this could be a
good opportunity.

✔ Nanoelectronics Group (http://nano.stanford.edu/about.php)
is all about graduate students exploring semiconductor technology,
solid-state devices, and electronic imaging. Their list of visiting schol-
ars is impressive, and getting accepted into this group is a solid career
builder.

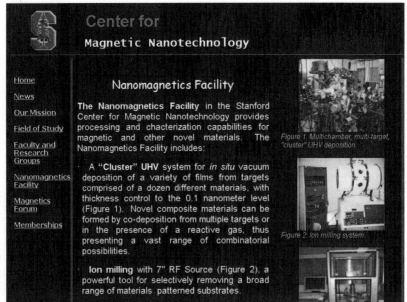

Figure 16-9:
Stanford's
Center for
Magnetic
Technology
web site.

For a more extensive list of colleges and universities offering education in
nanotechnology, go to www.understandingnano.com/nanotechnology-
university-college-education.html.

Chapter 17

Ten Interesting Nano Research Labs

In This Chapter

▶ Exploring labs with interesting specialties

▶ Surveying labs in different countries

▶ Scoping out a corporate lab

▶ Discovering labs with government affiliations

So much activity is going on in the area of nanotechnology research that assembling a list of ten labs was an interesting task. Should we presume to pronounce the ten best labs? Should we rank them by how much money is being poured into their programs? Or should we choose them by their list of accomplishments?

In the end, we decided to provide you with a cross section of focuses (for example, medicine and energy), a variety of countries, and different types of organizations (from government funded labs to those run by universities or corporations).

With apologies to any labs we couldn't include in this short list, we think this chapter gives you an overview of some interesting research going on in the world of nano.

NanoTumor Center

www.ntc-ccne.org

The NanoTumor Center was established in 2005. In that year, the National Cancer Institute (NCI) started a $144 million effort involving eight cooperative programs with universities in the United States. These programs were named Centers for Cancer Nanotechnology Excellence (CCNE). Six universities were brought into a consortium led by the University of California at San Diego to create the NanoTumor Center. Each center involved in this program must bring resources from various disciplines, such as engineering, chemistry, physics, and health sciences, to attack the problem of fighting cancer by using nanotechnology.

The NanoTumor Center has stated that its primary focus is "to apply nanotechnology to the treatment, understanding, and monitoring of cancer towards reducing the suffering and death it results in. To realize this objective, we use targeted nanoparticles of various sizes and properties, optimized for detection, sensing, imaging, and therapy."

One area of focus is to find a way to insert nanoparticles into the blood without the body's immune system detecting them. After nanoparticles enter the blood system undetected, they could attach to a tumor and penetrate it without causing damage to surrounding organs. These particles could then provide measurements of a tumor's growth to help researchers understand more about cancer. The Center's intention is to then communicate what they find to corporations, who could develop products and procedures to use in a clinical setting.

Longer term, the lab hopes to develop nanoplatforms, which they call "a payload of multifunctional smart motherships." These micron-sized motherships would deliver a collection of nanoparticles that could detect and identify tumors, and also provide images and measurements to researchers or medical practitioners. They could also deliver treatment to destroy cancer cells as they travel through the blood system. The team working on this effort includes doctors, scientists, mathematicians, and engineers, as well as business people whose companies could collaborate with the lab to offer these advanced treatments to patients.

Here are two of the more interesting accomplishments of the NanoTumor Centers:

- Dr. Thomas Kipps has developed a chemically engineered adenovirus nanoparticle. This virus can be used to deliver a molecule that affects the immune system, alerting it to an infection in the body. In tests, researchers have seen that only one injection of this nanoparticle-based virus has resulted in significant reductions in the number of leukemia cells in some patients and in the size of tumors in the lymph nodes and spleen in others. One patient went into remission after treatment with this engineered virus.

- Researchers at the NanoTumor Center, along with other organizations, such as MIT–Harvard Center of Cancer Nanotechnology, Excellence, have helped to develop magnetic nanoworms. These chains of iron oxide nanoparticles make it possible to improve magnetic resonance imaging (MRI) by enhancing the contrast in the images. This program has also created a method for using gold nanorods to heat a tumor so that proteins rise to the surface. Other nanoparticles can then more easily attach themselves to the tumor, making it possible to take higher resolution images of the tumor. This capability to change the properties of cancerous tissue to improve diagnosis is a very promising area for cancer research.

You can find out more about these accomplishments at the NanoTumor Center's web site, which is shown in Figure 17-1.

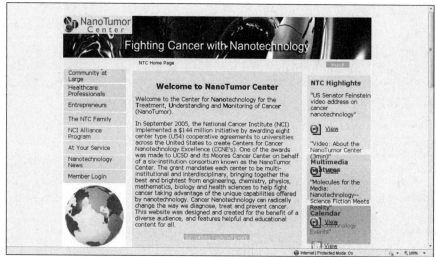

Figure 17-1:
NanoTumor
Center web
site.

The Molecular Foundry

www.foundry.lbl.gov

The Molecular Foundry is part of Lawrence Berkeley National Laboratory, which is funded by the U.S. Department of Energy. Started in 2006, this lab's stated primary focus is "providing support to researchers from around the world whose work can benefit from or contribute to nanoscience." The lab offers this support in the form of access to expensive instruments, nanomaterials, and researchers with hefty credentials and expertise. According to their web site (see Figure 17-2), their hope is that by making these resources available, they will help further "the synthesis, characterization, and theory of nanoscale materials."

A so-called Scientific User Facility, this lab makes it possible for researchers to run tests and examine information that they might not have access to on their own. This access is global: Chemists, physicists, engineers, and others from around the world use these facilities, after their proposal has been approved by a review board. Proposals are accepted twice a year and are approved based on their potential to get the best benefit out of using the Molecular Foundry's resources. Use of the resources is free if an individual or organization agrees to keep their research in the public domain and publish their findings. Those who want to keep their data confidential must pay a fee.

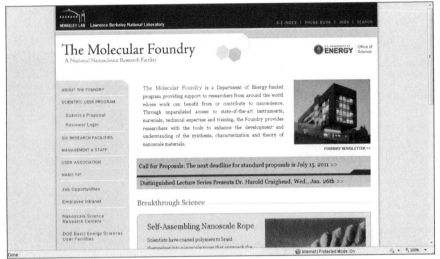

Figure 17-2:
The
Molecular
Foundry
web site.

The Foundry is a collection of six facilities. Beyond making their facilities available to others, they also perform their own research based on the following four themes:

- **Combinatorial nanoscience:** This research focuses on the use of robotic synthesizers to obtain information about biological and inorganic nanostructures. These synthesizers can test thousands of compounds at one time, so they can quickly accumulate data about the properties of these structures.

- **Nanointerfaces:** This area involves combining different types of nanomaterials, for example, inorganic nanoparticles with biological molecules, to create hybrid nanomaterials. These hybrid nanomaterials have special properties such as greater strength.

- **Multimodal in situ nanoimaging:** Researchers with this focus use imaging techniques to investigate the interaction between nanoparticles in liquids and vapors.

- **Single-digit nanofabrication:** This area of research involves building structures that have features less than 10 nanometers wide. One application of advances in this area will be the capability to pack more computing power or memory into integrated circuits on computer chips.

Each of the six facilities has its own focus, which falls into one of these areas:

- Imaging and manipulation of nanostructures

- Nanofabrication, especially advanced lithographic and thin-film processing

- Inorganic nanostructures, including semiconductor, carbon, and hybrid nanostructures

✔ Organic and macromolecular synthesis, for example synthesizing organic molecules into materials

✔ Biological nanostructures

✔ Theory of nanostructured materials used to help understand new principles and behaviors in nanoscale materials

Some of the Molecular Factory's accomplishments to date include

✔ Creation of nanocrystals that help to investigate activity in cells: These nanocrystal probes emit light to help researchers study components of living cells and other complex systems.

✔ A low-cost method of producing solar cells: Using nanocrystals, a startup company called Solexant Corporation is manufacturing lower-cost solar cells for use in flexible solar panels. Solexant is aiming to produce solar cells that cost less to manufacture and are very effective in converting sunlight to electricity. We discuss how nanotechnology is improving solar cells in more detail in Chapter 10.

The Nanomedicine Center for Nucleoprotein Machines

www.nucleoproteinmachines.org

Most of the medical world focuses on treating symptoms of diseases, not the underlying genetic causes. The Nanomedicine Center for Nucleoprotein Machines, started in 2006, is one organization that is focused entirely on the genetic causes of disease. Their hope is that we can one day treat many common diseases directly at the genetic level.

The Center is one of six nanomedicine development centers funded by the U.S. National Institutes of Health. Participating institutions are Georgia Institute of Technology, Stanford University, New York University, Cold Spring Harbor Laboratory, Emory University, Harvard Medical School, Medical College of Georgia, and MIT.

This Center is interesting in that it is focused on one challenging issue, which they state as "understanding and re-directing natural processes for repair of damaged DNA." The Center's five-year goal is to reengineer the "homologous recombination repair machine" to provide a clinically applicable gene correction technology."

The *Nucleoprotein Machines* part of the Center's name relates to the fact that proteins are nature's workhorses for repairing damaged DNA. A nucleoprotein is simply a protein that is attached to a nucleic acid such as DNA.

Homologous recombination is a form of genetic repair. When a break occurs in DNA, double-strand homologous recombination is a natural process in which the structure of a neighboring DNA is copied onto the damaged DNA to repair it. This process occurs constantly in our bodies to repair naturally occurring damage to DNA.

The current focus at the Center involves producing a gene correction device in the form of an engineered protein. This protein is designed to detect a defective DNA double strand and cut off the defective portion. After the DNA double strand has been cut, it can be repaired by the natural homologous recombination process.

The Center is testing its methods on sickle-cell disease in mice. Sickle-cell is a genetic disease that causes red blood cells to have an abnormal shape. This odd shape results in the cells moving through the bloodstream less easily than normally shaped red blood cells do, which results in anemia. Sickle-cell, which has no known cure, is a painful disease that shortens life. The Center uses mice and models of sickle-cell disease to find a way to repair the mutated gene that causes it. This focus on sickle-cell could lead to repair of other single-gene diseases.

According to the Center's 2010 Progress Report — Executive Summary, the areas of development are as follows:

✔ Development of nanoprobes to investigate the assembly, disassembly, and control of DNA repair machines. Machines in genes manage the storage of information about cells. The Center has a goal of learning how to modify the information stored in DNA and RNA. Initial work has begun, though they acknowledge that solving the puzzle could take decades.

✔ Development of new tools to create a model for nanomachines to assist in the repair of DNA double-strand breaks. Researchers will use these machines to understand the functions of DNA and eventually learn how to manipulate it.

✔ Development of strategies for tagging components of DNA and RNA.

✔ Synthesis of quantum dots (semiconductor nanocrystals) that are less bulky than those commercially available.

✔ Development of small beacons that offer the means to image protein interactions. The Center is the first to capture real-time images of certain proteins.

✔ Development of a method of delivering proteins to the nucleus of human cells.

You can visit the Center's web site, which is shown in Figure 17-3, for more information about their fascinating work.

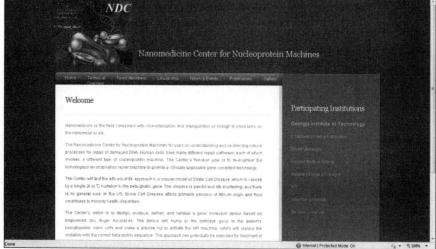

Figure 17-3:
The Nano-
medicine
Center for
Nucleo-
protein
Machines.

London Centre for Nanotechnology

www.london-nano.com

The London Centre for Nanotechnology (LCN) was founded in 2003 as a partnership between University College London and Imperial College London. LCN emphasizes that they are the only such facility located in the heart of a major city, which means that they can develop close ties to investment and industrial colleagues. They also use the resources of both colleges to offer nanotechnology training across disciplines, such as electronic and mechanical engineering, chemistry, and earth sciences. In addition to educating nanotechnologists, LCN seeks to educate the public about the potential of nanotechnology in our world.

LCN's purpose is to solve global problems in the three areas that they refer to as Grand Challenges because of their vital importance to the world:

- ✔ Healthcare
- ✔ Information technology
- ✔ Planetcare (climate change and energy sources)

LCN has had successes in several areas. For example, LCN is working to make nanotubes more useful by developing a way to activate sites on the nanotubes' surfaces to make it easier for molecules to bind to them. This work makes it possible to manufacture a variety of functionalized carbon nanotubes in large-scale production quantities. Manufacturers of carbon nanotubes could essentially create nanotubes with specific chemical properties, making them compatible with other materials or useful for initiating catalytic reactions. Because so many molecules can bond with nanotubes, this capability offers some exciting opportunities.

The process of modifying the properties of nanoparticles, such as nanotubes, by binding molecules to them is called functionalization and is discussed in more detail in Chapter 4.

Another accomplishment at LCN is the speeding up of DNA sequencing. In recent experiments, they used nanotechnology and single molecules to enable researchers to count the number of nucleotides in a DNA strand. They essentially tunnel a DNA strand through a nanopore and record the current of the strand as it passes through. This procedure would help to uncover the sequence in the strand because the four DNA nucleotides tunnel in different ways.

A nucleotide is a molecule that forms the units that make up the structure of DNA and RNA.

The London Centre for Nanotechnology's web site is shown in Figure 17-4.

Figure 17-4:
London
Centre for
Nano-
technology
web site.

Cornell NanoScale Science & Technology Facility

`www.cnf.cornell.edu`

Located at Cornell University, the Cornell NanoScale Science & Technology Facility (CNF) is part of the National Nanotechnology Infrastructure Network (NNIN), which in turn is funded by the National Science Foundation. Like the Molecular Foundry, CNF offers user facilities to those who want to advance nanotechnology. Unlike the Molecular Foundry, however, the process to get access to their facilities is less arduous. Both academic and industry researchers can use their instruments, and experts to perform their own experiments. Researchers who want to use the equipment are charged a fee that varies depending on whether the user is from academia or industry.

For a listing of all NNIN labs and links to their web sites, see the Nanotechnology Research page of our companion web site at `www.understandingnano.com/nanotechnology-research.html`.

The main focus of the folks at CNF is to "support a broad range of nanoscale science and technology projects by providing state-of-the-art resources coupled with expert staff support. Research at CNF encompasses physical sciences, engineering, and life sciences, and has a strong inter-disciplinary emphasis."

Each year, CNF attracts more than 700 users who spend an average of a week or two at their facilities. Their resources are so heavily used that they are open 24 hours a day.

CNF, whose web site is shown in Figure 17-5, isn't only a landlord. Of the work undertaken in their facilities, 50 percent is their own. Following are a few areas of research that were included in their report of accomplishments for the 2009–2010 year (a 267-page report):

- Vertically aligned carbon nanotube membranes for solar hydrogen production: This research holds the potential for creating a solar-powered method of producing hydrogen for applications such as fuel cells, without adding carbon dioxide pollution to our air.

- Transfer-free fabrication of graphene transistors: Transistors made with graphene can handle much higher frequency signals than transistors made with silicon. This research offers the possibility of providing a way to mass produce transistors and integrated circuits made of graphene.

- On-chip spectrophotometry for bioanalysis using nanophotonic devices. This research is aimed at developing low-cost, automated diagnostics systems for medical applications.

Figure 17-5:
Web site of
the Cornell
NanoScale
Science &
Technology
Facility.

HP Labs

`www.hpl.hp.com/research/about/nanotechnology.html`

Do some corporations perform internal nano research? You bet they do. One example is HP Labs, part of Hewlett Packard, the printer and computer giant. Their lab was set up in 1966 when the founders Hewlett and Packard decided that having their own lab would help their researchers focus on technology that could shape the future of the company. Building on past successes with technology such as scientific calculators, cordless mice, and thermal inkjet printing, the lab today works in a variety of areas, including nanotechnology. Their main nano focus is on nanoelectronics, the use of nanotechnology to build smaller integrated circuits and streamline information processing.

HP Labs represents its focus this way: "We believe we have a practical, comprehensive strategy for moving computing beyond conventional silicon electronics to the world of molecular-scale electronics. We are investigating the underlying science of nanostructures that operate at the atomic scale, looking for advantageous ways of exploiting their unique properties."

One of their important advances in nano was the creation of the first molecular logic gate. Logic gates are parts of integrated circuits. A molecular logic gate is very small and can be used to chemically assemble an electronic nanocomputer — otherwise known as a microprocessor.

We get into nanoelectronics such as microprocessors in more detail in Chapter 6.

Nanoimprint lithography, which involves printing patterns on integrated circuits, was first used by HP Labs. This kind of lithography can be used to build very small electronic devices, such as a transistor with feature sizes as small as 12.5 nanometers. This size device could not be built using the more traditional optical lithography because light cannot be used to print features below a certain size.

A recent development from HP Labs that could prove significant in the area of memory density is the nanoscale memristor, which is an element of an electronic circuit. Although people had guessed at the existence of memristors, they were unable to prove their existence using larger scale devices. In 2008 HP Labs was able to demonstrate the memristor effect by using nanoscale devices because the electrical effect is stronger at that size. In large devices, the electrical effect from memristors had simply been viewed as noise.

The memristor is in the same class of circuit components as those basic electronics components — the resistor, capacitor, and inductor. What is special about the memristor is that its resistance depends on the voltage that was last applied to it, even after the voltage is removed. Therefore, a memristor can be used as a single-component memory cell in an integrated circuit, allowing higher memory density than conventional memory circuits.

HP has announced that it has made an agreement with a semiconductor company to use memristors on chips that will be sold as a commercial product. This advance could increase the memory capacity of computer chips by as much as a factor of 10. The new materials created in this joint venture will be called ReRAM (resistive random access memory) and these materials have the potential of replacing flash memory currently used in mobile phones, MP3 players, and even computer hard drives.

Memristors take less energy and are faster than current memory storage technologies. They can even retain memory after you turn off the power to a device.

The global nature of the activities of HP Labs is evidenced on their web site, shown in Figure 17-6.

Figure 17-6:
Web site
of HP Labs
lists their
global labs.

Center for the Environmental Implications of Nanotechnology

www.ceint.duke.edu

The Center for the Environmental Implications of Nanotechnology (CEINT) is funded by the National Science Foundation and the Environmental Protection Agency, which jointly came up with $14.4 million in seed money for the center in 2008. CEINT's primary focus is to look into the possibility that nanoparticles could themselves cause damage to our environment.

As you discover in various chapters of this book, nanoparticles are not only tiny but have different properties from the same material at the above-nano scale. This makes nanomaterials appealing because they can be used in a variety of settings to make products stronger, more lightweight, and so on. However, new materials with new properties could cause problems we can't anticipate. For example, silver nanoparticles are useful in killing bacteria and are already being used in some household products. One study by researchers at Purdue University, however, found that silver nanoparticles suspended in a solution were toxic to minnows. If silver nanoparticles released by detergents or other household products were to get into our water supply, they could pose a danger to fish and other life-forms.

The goal of CEINT is to establish the relationship between nanomaterials and possible side effects of their use, whether by creating byproducts, biological changes, or environmental damage. Four core universities are involved in the effort: Duke, Howard, Virginia Tech, and Carnegie Mellon. They share facilities for analysis and nanofabrication and jointly offer one of the best resources in the world for research in this area.

CEINT research revolves around three themes:

- ✔ Exposure: Transport and transformations in laboratory systems
- ✔ Cellular and organismal responses
- ✔ Ecosystem responses

Their work involves the creation of highly controlled ecosystems (also called mesocosms). These 3-by-12-foot areas host nanoparticles that the researchers study for their effects on organic life forms such as fish or bacteria. These mesocosms provide information used in experiments across the four facilities. You can visit their web site (shown in Figure 17-7) to learn more about their developments.

Figure 17-7:
Center for the Environmental Implications of Nanotechnology web site.

Behind the scenes at a nano lab

In 2008 coauthor Earl Boysen toured the NanoTech User Facility at the University of Washington in Seattle. This facility is part of the National Nanotechnology Infrastructure Network (NNIN), whose goal is to support nanotechnology research. Researchers can use the lab and its sophisticated instruments at a surprisingly low cost; the hourly price tag for using the lab is below the cost of the service contracts required to maintain the instruments. The facility makes up the difference from their funding through the National Science Foundation and the University of Washington.

Although Earl was lucky enough when he worked in the semiconductor industry to have access to this level of equipment at his company, not every company can afford such an investment. For them, the lab provides a cost-effective way to pursue nano research. The instruments that industrial or academic researchers can take advantage of include the following:

- Transmission electron microscope (TEM) with a resolution of 1.5 Ångstrom

- Scanning electron microscope (SEM) with an E-beam lithography system that can generate patterns with feature sizes as small as 20 nanometers

- Atomic force microscopes (AFM)

- Confocal Raman microscope

- Laser scanning confocal microscope

The facility provides training to qualify researchers on the equipment. If you want to send a few researchers to the NanoTech User Facility for an extended time to run their own tests, you can get office space free of charge. If you're not a hands-on type, the lab will have their staff run your tests for you. For example, you send them a sample, their staff runs tests, and the output of the SEM is then displayed to you online. This capability allows you to make suggestions to the staff members running the SEM, so they can optimize their analysis of the sample in real time.

Each lab in the National Nanotechnology Infrastructure Network has its own area of focus. The Nanotech User Facility takes advantage of expertise at the University of Washington to focus on applications of nanotechnology in biology. This emphasis on biology is evident in some of the capabilities at the NanoTech User Facility. For example, they demonstrated how the laser confocal microscope provides optical photos of the inside of a cell, which can be used to determine whether quantum dots have penetrated into the cell. (They deduce this from the location of the dots' emissions.) Another interesting demonstration of their biological focus was their capability to make arrays of square petrie dishes small enough to allow testing of individual cells.

Those interested in finding out more about the University of Washington's Nanotech User Facility can visit their web site at www.depts.washington.edu/ntuf/.

ARC Centre of Excellence for Functional Nanomaterials

www.arccfn.org.au

Down under in Queensland, Australia, the Australian Research Council has established the ARC Centre of Excellence for Functional Nanomaterials (ARCCFN), whose web site is shown in Figure 17-8. With headquarters at The University of Queensland, the lab has "nodes" at the University of New South Wales, Deakin University, and the University of Western Sydney.

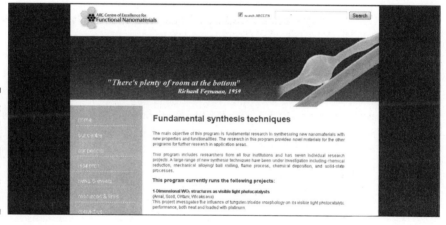

Figure 17-8: ARC Centre of Excellence for Functional Nanomaterials web site.

Starting with a grant in 2003, ARCCFN is part of the ARC centers of excellence initiative, which supports applied research and training in the sciences through a national competition. The primary focus of this group is "the novel synthesis, characterization, and applications of functional nanomaterials such as nanoparticles, nanotubes, thin films, and nanoporous and nanocomposite materials." They have a special focus on nanotechnology efforts that will have a serious effect on industries with a focus that is most important to the Australian economy, such as microelectronics, manufacturing, energy, medicine, and the environment.

The Centre runs 48 projects that include the following core programs:

- ✔ Fundamental synthesis techniques
- ✔ Computational nanomaterials science
- ✔ Clean energy production and utilization
- ✔ Environmental technologies
- ✔ Healthcare

Some of the developments of the Centre have included the following:

✔ A more cost-effective water purification method: This method is used to clean up industrial waste in wastewater streams that have low levels of organic pollutants so the water can be reused. Methods that use standard energy sources are expensive, but this method is more cost effective because it uses a photocatalyst made with titanium dioxide nanomaterials that help break down organic pollutants using sunlight as the power source.

✔ Improved targeted drug delivery: This method uses hybrid nanoparticles coated with a porous silica that enables them to be targeted for delivery to specific cells. According to this project's leader, Dr. Xu, "to be an effective delivery system, the nanoparticle must be a suitable carrier and readily transported through various biological barriers to the site of action."

Center for Atomic Level Catalyst Design

www.efrc.lsu.edu

Catalysts are used to reduce the energy required to make a chemical reaction. For example, your car has a catalytic converter that contains the catalyst platinum, which changes air pollution into harmless elements using much less energy than would otherwise be needed. Catalysts are vital to just about every energy-producing method available to us.

The Center for Atomic Level Catalyst Design is based at Louisiana State University, with researchers located at Clemson University, University of Florida, Georgia Tech, Grambling University, Louisiana Tech, Oak Ridge National Laboratory, Pennsylvania State University, Texas A&M University, Vienna University of Technology, and the University of Utrecht.

The Center was started in 2009 when the Office of Basic Energy Sciences in the U.S. Department of Energy set up 41 Energy Frontier Research Centers. As one of these Centers, the Center for Atomic Level Catalyst Design works with other universities, national labs, nonprofits, and corporations to solve what they refer to as grand challenges.

According to the mission statement for these 41 EFRCs, they "will harness the most basic and advanced discovery research in a concerted effort to establish the scientific foundation for a fundamentally new U.S. energy economy. The outcome will decisively enhance U.S. energy security and protect the global environment in the century ahead."

The Center has several major goals in the area of *nanocatalysts,* including the following:

✔ Improving the ability to simulate catalytic reactions

✔ Advancing the quality of tools available for identifying the characteristics of catalysts

✔ Developing tools for manufacturing new catalysts

The lab, whose web site is shown in Figure 17-9, has a team of investigators from various universities. Small interdisciplinary teams focus on one catalyst type.

Figure 17-9:
Center for Atomic Level Catalyst Design web site.

The Center is attempting to advance the science of making new catalysts by using a computer model that could predict the composition and structure of new catalysts required for specific chemical reactions. When a new catalyst is requested the catalytic material could then be produced with each atom arranged in the optimum structure to enable that particular chemical reaction.

California NanoSystems Institute

www.cnsi.ucla.edu

The UCLA research center known as CNSI was set up in 2000 as part of a California initiative that created four Institutes for Science and Innovation. One of their mandates was to create relationships with representatives of industry to find ways to advance technology to benefit society and the citizens of the State of California in particular. This effort is interdisciplinary, stressing collaboration among experts in life and physical sciences, engineering, and medicine.

The stated mission of CNSI is "to encourage university collaboration with industry and to enable the rapid commercialization of discoveries in nanosystems." To that end, CNSIs nanosystems research has four areas of focus:

- ✔ Energy
- ✔ Environment
- ✔ Health medicine
- ✔ Information technology

CNSI, whose web site is shown in Figure 17-10, includes eight core facilities that boast impressive laboratories, equipment such as electron microscopes and high throughput robotics, and clean rooms. These resources are available to CNSI, other UCLA faculty, and other representatives of the academic and industrial world.

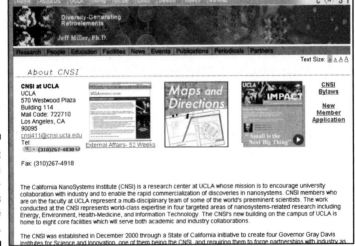

Figure 17-10: California Nano-Systems Institute web site.

Here's the menu of the resources that CNSI makes available: "both wet and dry laboratories, equipment in the form of electron microscopes, atomic force microscopes, X-ray diffractometers, optical microscopes and spectroscopes, high throughput robotics and class 100 and 1000 clean rooms."

CNSI's corporate associations are vital to their mission. These include $NanoH_2O$, a provider of membranes for reverse osmosis, BASF, which is focused on alternative energy sources, and Abraxis Bioscience, a biopharmaceutical company. These three companies are producing some intriguing research and development directions.

NanoH$_2$O is using nanotechnology to reduce the cost of desalination. They received four awards in 2010 for their work. NanoH$_2$O is already commercializing a desalination technology developed by a CNSI researcher, Eric Hoek. They incubated Hoek's idea at CNSI for two years before transferring the work to their own facilities.

BASF is using the nanopores in metal organic frameworks (MOFs) to bring us closer to the reality of natural gas–powered cars by increasing the amount of natural gas that can be stored in a tank. These cars could travel twice as far as current vehicles on a single tank of gas because the metal organic frameworks would help to improve gas storage capacities. BASF has worked with UCLA's Omar Yaghi, the person who discovered MOFs. They are focused on synthesis of MOFs on an industrial level to provide enough of this material to sell commercially for use in natural gas tanks and other applications.

Abraxis BioScience (now part of Celgene) is a company with a history of working in nanotechnology with its nanoparticle albumin-bound (nab) platform. Albumin is a protein that delivers nutrients to cells. Abraxis BioScience attaches chemotherapy drugs to albumin, allowing the drug to be delivered to cancer tumor cells. They were the first to get a drug product approved by the FDA using nab to treat breast cancer. The company is working with CNSI to develop new technologies in the world of nanomedicine. The company is contributing $10 million to fund projects with CNSI. Focusing on pharmaceuticals and biotechnology, they are working on improving diagnosis and treatment for diseases that are debilitating or life-threatening.

Glossary

Aerogels: An insulating foam of silica nanoparticles separated by nanopores containing air.

Armchair quantum wire: An electrically conducting wire constructed of metallic carbon nanotubes.

Atomic force microscope (AFM): A scanning probe device producing extremely high-resolution images of surfaces at the atomic scale.

Autonomous nanotechnology swarms (ANTS): Collections of very small robots that can collectively change their shape, communicate with each other, and act as sensors.

Bionanorobot: Robots approximately the size of the cells in our bodies that can perform tasks on nanoscale objects. The term *bio* refers to the fact that biological molecules are part of the nanorobot mechanism.

Bottom up: An approach to manufacturing that uses nanotechnology to build structures atom by atom.

Buckyball: Also called Fullerene. Molecules composed of carbon atoms arranged on the surface of a nearly spherical shape, in a pattern of pentagons and hexagons that looks like a soccer ball. Buckyballs are named after Buckminster Fuller, who introduced the geodesic dome. C60, indicating the number of carbon atoms in a single sphere, is the most common type of buckyball.

Carbon nanotubes: Molecules composed of carbon atoms arranged on the surface of a cylindrical shape in a pattern of hexagons as seen in buckyballs.

Covalent bonding: A type of chemical bonding in which electrons are shared between atoms.

Diamondoid: A very strong structure consisting of carbon atoms in a three-dimensional lattice joined by covalent bonds.

Dip-pen nanolithography: A technique for applying nano-sized patterns on surfaces using the tip of an atomic force microscope.

E-beam nanolithography: A technique for applying nano-sized patterns on surfaces using an electron beam.

Electron microscope: A type of microscope that uses a beam of electrons to produce an image.

Extreme ultraviolet nanolithography: A type of optical lithography that uses extremely short wavelengths of light.

Fullerene: *See* buckyball.

Functionalization: To attach sets of molecules to a nanoparticle to create a specific result, such as allowing a nanoparticle to bond to polymers to create a strong, lightweight composite material.

Graphene: Molecules composed of carbon atoms in a pattern of hexagons like that seen in buckyballs, arranged in a planar sheet one atom thick.

Mechanosynthesis: The use of mechanical tools to build a covalently bonded structure by depositing atoms or molecules at desired locations with atomic precision.

Metal-organic framework (MOF): Metal oxides linked by organic molecules in a porous crystalline structure that can be used to store gases.

Molecular assembler: *See* molecular fabricators.

Molecular fabricators: A machine that can create objects by positioning atoms or molecules to build a covalently bonded structure.

Molecular manufacturing: Building an object molecule by molecule.

Multi-walled nanotubes (MWNT): A nanotube that contains more than two layers with the smaller diameter nanotubes inside the larger diameter nanotubes.

Nanocatalyst: Nanoparticles of a substance, such as platinum, that reduce the temperature at which various chemical reactions occur.

Nanocomposite: A matrix to which nanoparticles have been added to improve a particular property of the material, such as strength.

Nanoelectromechanical systems (NEMS): Devices that integrate nanoscale mechanical and electrical components into a single component. NEMS are generally built on semiconductor wafers using integrated circuit manufacturing techniques.

Nanoelectronics: Electronics devices containing components such as sensors or transistors that have features less than 100 nanometers in size.

Nanoengineering: The study and practice of engineering at the nanoscale.

Nanofilm: A film containing nanoparticles that modify the properties of the film.

Nanolithography: The practice of printing nanoscale patterns on a surface.

Nanomaterials: Materials with one or more dimensions measuring less than 100 nanometers.

Nanomedicine: The use of nanotechnology for diagnosing, treating, and preventing disease.

Nanometer: A measurement equal to 10^{-9} meter, or 1 billionth of a meter. Approximately 800 100-nanometer particles placed side by side would match the width of a human hair.

Nanoparticle: Particles ranging from 1 to 100 nanometers in diameter.

Nanoparticle field extraction thruster (nanoFET): A device in which nanoparticles are charged by losing electrons when they touch an electrode at a positive voltage. After the nanoparticles are charged, an electric field can accelerate them, providing thrust to a spacecraft.

Nanorobot: Robots approximately the size of the cells in our bodies that have a propulsion system, sensors, manipulators, and possibly an on-board computer that can perform tasks on nanoscale objects.

Nanorod: A nanoscale object in which all dimensions are between 1 and 100 nanometers in size with the length typically three to five times the diameter. Nanorods are typically synthesized from metal or semiconducting materials.

Nanoscience: *See* nanotechnology.

Nanosensor: Sensors that use nanoscale materials to detect biological or chemical molecules.

Nanoshells: Nanoparticles consisting of a gold coating over a silica core. Nanoshells can convert infrared light into heat to destroy cancer cells.

Nanotechnology: The study and use of structures between 1 nanometer and 100 nanometers in size.

Nanotube: *See* carbon nanotube.

Nanowire: A wire of any length with a diameter of less than 100 nm.

National Nanotechnology Infrastructure Network: A partnership of 13 nano-technology user facilities funded by the National Science Foundation.

National Nanotechnology Initiative: A United States government agency created by the 21st Century Nanotechnology R&D Act tasked with coordinating the efforts of several member agencies working on developing nanotechnology.

Quantum dot: A semiconductor nanoparticle that emits light when irradiated by ultraviolet light. The color of the light emitted depends on the size of the quantum dot.

Scanning electron microscope (SEM): A device used to obtain images of nanoscale surface details using an electron beam.

Scanning tunneling microscope (STM): A device that provides images of molecules and atoms on conductive surfaces.

Self-assembly: A technique for using the chemical bonding and repulsive properties of functionalized nanoparticles with other atoms or molecules to cause them to assemble in regular patterns without external intervention.

Single-walled carbon nanotube (SWNT): *See* carbon nanotube.

Targeted drug delivery: The use of nanotechnology to deliver drugs directly to diseased cells within the body.

Top down: An approach to manufacturing that removes portions of bulk materials to create nano-sized features.

Transmission electron microscope (TEM): An electron microscope device in which electrons are sent through a sample to produce an image at sufficient magnification to show the position of the atoms in the sample.

Index

• *O* •

• T •

Apple & Macs

iPad For Dummies
978-0-470-58027-1

iPhone For Dummies,
4th Edition
978-0-470-87870-5

MacBook For Dummies, 3rd
Edition
978-0-470-76918-8

Mac OS X Snow Leopard For
Dummies
978-0-470-43543-4

Business

Bookkeeping For Dummies
978-0-7645-9848-7

Job Interviews
For Dummies,
3rd Edition
978-0-470-17748-8

Resumes For Dummies,
5th Edition
978-0-470-08037-5

Starting an
Online Business
For Dummies,
6th Edition
978-0-470-60210-2

Stock Investing
For Dummies,
3rd Edition
978-0-470-40114-9

Successful
Time Management
For Dummies
978-0-470-29034-7

Computer Hardware

BlackBerry
For Dummies,
4th Edition
978-0-470-60700-8

Computers For Seniors
For Dummies,
2nd Edition
978-0-470-53483-0

PCs For Dummies,
Windows
7th Edition
978-0-470-46542-4

Laptops For Dummies,
4th Edition
978-0-470-57829-2

Cooking & Entertaining

Cooking Basics
For Dummies,
3rd Edition
978-0-7645-7206-7

Wine For Dummies,
4th Edition
978-0-470-04579-4

Diet & Nutrition

Dieting For Dummies,
2nd Edition
978-0-7645-4149-0

Nutrition For Dummies,
4th Edition
978-0-471-79868-2

Weight Training
For Dummies,
3rd Edition
978-0-471-76845-6

Digital Photography

Digital SLR Cameras &
Photography For Dummies,
3rd Edition
978-0-470-46606-3

Photoshop Elements 8
For Dummies
978-0-470-52967-6

Gardening

Gardening Basics
For Dummies
978-0-470-03749-2

Organic Gardening
For Dummies,
2nd Edition
978-0-470-43067-5

Green/Sustainable

Raising Chickens
For Dummies
978-0-470-46544-8

Green Cleaning
For Dummies
978-0-470-39106-8

Health

Diabetes For Dummies,
3rd Edition
978-0-470-27086-8

Food Allergies
For Dummies
978-0-470-09584-3

Living Gluten-Free
For Dummies,
2nd Edition
978-0-470-58589-4

Hobbies/General

Chess For Dummies,
2nd Edition
978-0-7645-8404-6

Drawing
Cartoons & Comics
For Dummies
978-0-470-42683-8

Knitting For Dummies,
2nd Edition
978-0-470-28747-7

Organizing
For Dummies
978-0-7645-5300-4

Su Doku For Dummies
978-0-470-01892-7

Home Improvement

Home Maintenance
For Dummies,
2nd Edition
978-0-470-43063-7

Home Theater
For Dummies,
3rd Edition
978-0-470-41189-6

Living the
Country Lifestyle
All-in-One
For Dummies
978-0-470-43061-3

Solar Power Your Home
For Dummies,
2nd Edition
978-0-470-59678-4

Available wherever books are sold. For more information or to order direct: U.S. customers visit www.dummies.com or call 1-877-762-2974.
U.K. customers visit www.wileyeurope.com or call (0) 1243 843291. Canadian customers visit www.wiley.ca or call 1-800-567-4797.

Internet

Blogging For Dummies,
3rd Edition
978-0-470-61996-4

eBay For Dummies,
6th Edition
978-0-470-49741-8

Facebook For Dummies,
3rd Edition
978-0-470-87804-0

Web Marketing
For Dummies,
2nd Edition
978-0-470-37181-7

WordPress
For Dummies,
3rd Edition
978-0-470-59274-8

Language & Foreign Language

French For Dummies
978-0-7645-5193-2

Italian Phrases
For Dummies
978-0-7645-7203-6

Spanish For Dummies,
2nd Edition
978-0-470-87855-2

Spanish
For Dummies,
Audio Set
978-0-470-09585-0

Math & Science

Algebra I
For Dummies,
2nd Edition
978-0-470-55964-2

Biology For Dummies,
2nd Edition
978-0-470-59875-7

Calculus For Dummies
978-0-7645-2498-1

Chemistry For Dummies
978-0-7645-5430-8

Microsoft Office

Excel 2010 For Dummies
978-0-470-48953-6

Office 2010 All-in-One
For Dummies
978-0-470-49748-7

Office 2010 For Dummies,
Book + DVD Bundle
978-0-470-62698-6

Word 2010 For Dummies
978-0-470-48772-3

Music

Guitar For Dummies,
2nd Edition
978-0-7645-9904-0

iPod & iTunes For
Dummies, 8th Edition
978-0-470-87871-2

Piano Exercises
For Dummies
978-0-470-38765-8

Parenting & Education

Parenting For Dummies,
2nd Edition
978-0-7645-5418-6

Type 1 Diabetes
For Dummies
978-0-470-17811-9

Pets

Cats For Dummies,
2nd Edition
978-0-7645-5275-5

Dog Training For Dummies,
3rd Edition
978-0-470-60029-0

Puppies For Dummies,
2nd Edition
978-0-470-03717-1

Religion & Inspiration

The Bible For Dummies
978-0-7645-5296-0

Catholicism For Dummies
978-0-7645-5391-2

Women in the Bible
For Dummies
978-0-7645-8475-6

Self-Help & Relationship

Anger Management
For Dummies
978-0-470-03715-7

Overcoming Anxiety
For Dummies,
2nd Edition
978-0-470-57441-6

Sports

Baseball
For Dummies,
3rd Edition
978-0-7645-7537-2

Basketball
For Dummies,
2nd Edition
978-0-7645-5248-9

Golf For Dummies,
3rd Edition
978-0-471-76871-5

Web Development

Web Design
All-in-One
For Dummies
978-0-470-41796-6

Web Sites
Do-It-Yourself
For Dummies,
2nd Edition
978-0-470-56520-9

Windows 7

Windows 7
For Dummies
978-0-470-49743-2

Windows 7
For Dummies,
Book + DVD Bundle
978-0-470-52398-8

Windows 7 All-in-One
For Dummies
978-0-470-48763-1

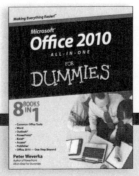

Wherever you are in life, Dummies makes it easier.

From fashion to Facebook®, wine to Windows®, and everything in between, Dummies makes it easier.

Visit us at Dummies.com

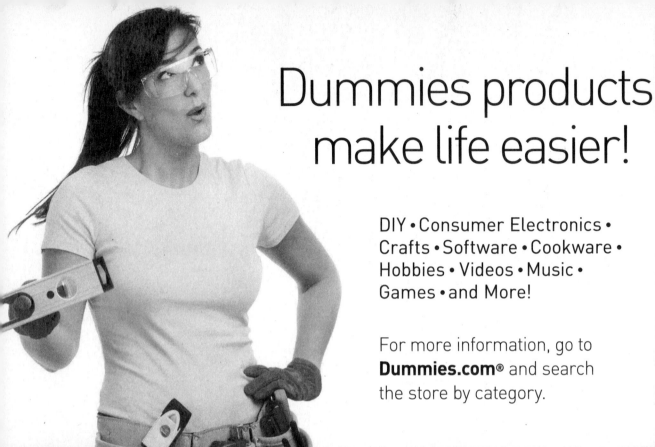

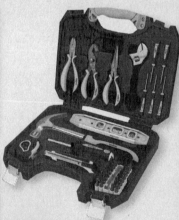